Plants-a-Plenty

Plants-a-Plenty

How to Multiply Outdoor and Indoor Plants
Through Cuttings, Crown and Root Divisions,
Grafting, Layering, and Seeds

by
Catharine Osgood Foster

Photographs by Rodale Press Photo Lab
Illustrations by D. Erick Ingraham

Rodale Press, Emmaus PA

Printed on recycled paper

Library of Congress Cataloging in Publication Data

Foster, Catharine Osgood, 1907-
   Plants-a-plenty.

   Bibliography: p.
   Includes indexes.
   1. Plant propagation. I. Title.
SB119.F67   635'.04'3   77-507

               ISBN 0-87857-156-6

             2 4 6 8 10 9 7 5 3 1

Other books by the same author

THE ORGANIC GARDENER
TERRIFIC TOMATOES
ORGANIC FLOWER GARDENING

To Thomas Foster
for all his help
and patience

CONTENTS

Introduction

Part I Propagation Techniques

PART II Plants to Propagate

Introduction

Every summer I pick marigolds and make an arrangement for the dining room table by pushing the base of the stems down on a pinholder in a shallow dish of water. I use both single flowers on their stems and quite large pieces of green top growth to fill out the arrangement and make a nice design. These pieces stay green and fresh for quite a long time, so all I have to do to keep the arrangement going is insert new blossoms when I remove the marigolds which have faded.

I have never ceased being excited every summer on the day when I look down into the dish and see that the green pieces of marigold plant have sent out half-a-dozen roots into the surrounding water. If it is late enough in the summer I make immediate plans to put the rooted cuttings in pots to make house plants for the fall and early winter. Sometimes I just let them stay in the water, where they will thrive for a month, at least, if the water is kept clean with frequent flushing and bits of charcoal. Buds should be removed to help the plant put its vigor into making new roots, but I have found that many of the marigold cuttings I have made naturally give up making buds when they're in water for longer than a week or so.

Any kind of propagation can be exciting, whether it's a simple rooting process like that of the marigolds in water, a delicate, complicated grafting or budding job like those done by professional horticulturists, or the marvelous enterprise of breeding and cross-pollinating your fruit trees and berries or other plants in order to find good new varieties. And I believe that everyone who ever planted a seed as a child remembers the delight and even awe felt when the seed came up. I don't think we ever forget that feeling and excitement about newness; and I am convinced that I hear it each year in my husband's voice when he says one morning in early May, "The peas are up!" And his voice seems to have the ring of a child's when he says in late autumn, "Oh, look! It's snowing!" Or one day in the spring, "I found the first crocus!" or "I saw the first robin." The newness of life is ever new; and we never seem to become so jaded that we get over responding to it.

There are plenty of other reasons for practicing plant propagation, but I put excitement first because I believe in pleasure before business when it comes to plants, even though some people earn their living from them. The miracle of new plants from old is always there, no matter how businesslike you become. And perhaps living with this mysterious wonder is the reason why so many gardeners have been rather mellow characters, and why propagation is a phase of gardening particularly appealing to them.

Whether you garden outdoors or indoors, the best way to have a young, healthy, vigorous collection of plants is to practice various methods of propagation: getting new plants from stem-tip cuttings or leaf cuttings, from air layering if the plants are suitable for that, from separating rhizomes or bulbs, or from any of the many other methods described in the chapters which follow. Of course, everyone who grows annual vegetables and flowers from seeds is well acquainted with the vigor of young plants which come quickly to fruition or flower in one season. If you save seeds from the most vigorous of your own plants, and do this over and over, you can probably develop a strain that is adapted to your own growing conditions and more vigorous than the one you started with. It doesn't really pay to save hybrid seeds because they do not breed true (become like the parent plants). But others are certainly well worth trying.

Perhaps you plan to have fruit trees and want to learn how to graft them for greater variety, pest resistance, or for a change in size. The same goals have led the experts to develop methods of grafting to suit these purposes. With the method called top-grafting, or topworking, you can even change old mature trees into something new.

You can start your propagating easily, perhaps by cutting suckers away from a rugged old snake plant (*Sambucus sansevieria*). Or you can get into really difficult, precise processes like growing orchids from tiny, almost invisible seeds which need to be grown on a sterile medium under exact conditions of light and heat. You can sow tomato seeds in a tin can filled with garden soil or on a specially prepared medium in a costly indoor greenhouse fitted with heating cables and special lights. You can start cuttings from quickly responding plants like the marigolds described above, or from difficult plants, like rhododendrons, which need mist-spray or special treatments.

The plants you grow may be for your own benefit, for sale, or for the worthy cause of propagating wild flowers to put back in their natural habitats to help save them from extinction. Whatever your aims, you might as well learn about some of the methods that have proven most successful. The common method, of course, is planting seeds, which are the result of sexual reproduction in all cone-bearing and flowering plants. But you can also learn to grow ferns from spores. They and many seed-bearing plants can also be propagated asexually, or vegetatively, by those other methods. For example, many fruit trees, house plants, and hybrid ornamentals are grown from cuttings and graft unions in order to keep the strain pure and to get daughter plants which will definitely breed true to the parent plant. Asexual means are often much quicker, too.

I am glad that I have learned about the many plant propagation practices that have been developed over the years. The chapters which follow in Part I attempt to describe them, and Part II sets forth the names of various kinds of plants, with suggestions for the propagation of each and some other bits of advice. Along the way I have also learned some of the basic botanical facts which make propagation possible. These I have also tried to share in order to convey something of their importance to the practical gardener who probably would like to know a little about the processes that go on inside plants.

I am also glad to know some of the figures I have read about propagation; in fact, I find them simply astounding. If you start with half-a-dozen bean seeds, plant them, then save and sow every seed of that harvest, and keep doing this for six years or six harvests, you'd end up with about 10,000 pounds of beans. Commercial growers count on such returns, and I guess that's why seedsmen so rarely run out of seeds. On a smaller scale, if you clip your hedge several times a summer, you can gather hundreds of twigs to start as cuttings for future plants and future hedges. Even a little pruning of one of your lilacs in late summer, for instance, might yield you enough branches to provide 10-inch hardwood cuttings for the entire neighborhood. Single orchid plants can produce a million seeds a year, and the spores from which ferns propagate run into the millions, too. This capacity of nature to provide plants-a-plenty is wonderful.

But equally wonderful is the capacity for vegetative reproduction; each cell in a plant body carries within it the genetic code for every other part of the plant. This means that theoretically, each of these millions of cells has the potential to produce a whole plant. With numbers like this you'd think that there would be no endangered species, but in nature each plant is adapted to a specific habitat, and many factors—especially man and his bulldozings and other doings—are at work. And all too many of them are upsetting the patterns which nature has been evolving and protecting for eons.

I am glad that these patterns are so complex that they escape total discovery. There's always the fun of something more to learn and the excitement of new discovery. Indeed, the lives of plants are so intricately interwoven with the lives of soil microorganisms, pollinating insects, and seed-carrying birds and animals that all these interrelationships are not anywhere near being understood. But the lure of discovery and understanding isn't the only thing that turns people into plant propagators.

A very practical reason for making cuttings, for instance, is the need for more space, both outside when perennials and shrubs get too big or too crowded, and inside when some large, luxuriant house plant usurps a whole corner of a room or blocks out all the light. Air layering of the top of such a plant can make a new, more compact plant to pot up and use in place of the "overwhelmer." If a plant gets scraggly, you also have a good reason to cut it back and start new plants from the cuttings. It is only sensible to keep enough for adequate replacement, but you can always give away other rooted cuttings which you are not going to need. And if you have entirely too many cutoff pieces, there is the compost heap which is forever in need of more organic materials.

Our garden club has a greenhouse committee and the use of a small greenhouse for experimental purposes. The main purpose of this committee during part of the year is to grow enough plants for the yearly plant sale that helps the club to make ends meet and fill up its coffers a little. So the first consideration is to grow plants that sell well. But the eager young horticulturists who lurk there also grow plants that they know little about just to see what will happen. The first year they had the greenhouse they were so excited, so absorbed in getting used to all the arrangements and equipment, and so busy scrounging around for plant materials and containers that they forgot to keep a day-by-day diary of all observations and events. But in time the diary will be started and passing on lore to future committee members will become an important tradition.

We can all be glad for the records kept by professional growers and amateurs, too; and for all the published accounts of what happens when you practice the various arts of propagating. My thanks go to all the authors I have read; most of them are listed in the bibliography at the end of this account. I thank all who have told me about their experiences and given me hints about helpful things to know. And I especially thank my husband Tom for putting up with all sorts of odd things growing in our kitchen, and for complaining only very minimally about the plants blocking the view from the· south windows and cluttering up our porch in the summer. We both love plants, however, and love to have them around. Like me, Tom finds plants endlessly interesting to watch and learn about, though I *have* heard him grumble about the "boring" avocado trees in the kitchen and also about the mice which ate some of the Christmas cactus when it was resting in the cellar one fall. (I expected him to grumble about that, but I was really surprised to hear that he felt so strongly about avocados!)

And so, of all the reasons I can think of for propagating plants I come back to pleasure and excitement. You'll find that most of the skills involved are not hard to learn, and that the enjoyment and entertainment you can get are boundless.

PART I

Propagation Techniques

Chapter 1

Things to Know and
Things to Have

For the early steps in propagating you will need certain materials and equipment that have been found successful over the years for various methods and various kinds of plants by gardeners and professional breeders. Even so, your options are many.

I wish I could say that there was one positive, best, or one-and-only way to make a mixture, for starting cuttings, for example, but I cannot. Many starting and potting mixtures have succeeded, so you might as well try out what appeals to you and what is possible with the supplies and equipment you find easiest to get. But there are important principles to keep in mind when you are experimenting.

Planting Mediums

For cuttings, the starting medium should be clean, or sterile, sharp, and rather scratchy, with small enough particles so that the mixture can be packed down to come in contact with the stem you insert. It must be able to hold water and to circulate some air. Make sure at least one ingredient is included which can also absorb water, but there should not be too much of

*Starting Mixtures for
Cuttings and Seeds*

this because it might block good drainage and the entrance of enough air for the young roots which will come from the cutting in time. A soggy medium for stems like those of geraniums will encourage black rot and ruin the cuttings.

The scratchiness is desirable so that there will be extra little wounds made along the stem, all of which raise your chances of getting the response that will cause good rooting. If, however, you choose a mixture that is not very scratchy to start your cuttings, have more sand on hand to line the hole where you insert the cutting so it can be scratched when you put it in.

Several recipes for starting or rooting mixtures are given in the next chapter, but one common proportion of materials used by many experts is a 1-1-1 combination of coarse builder's sand, to help drainage and to be scratchy; vermiculite or perlite; and milled sphagnum moss or peat moss, to support the stems and to hold water. You can put the moss through a sieve to shred it; and after you do this, moisten it slowly, let it rest for several hours, and then press it down before measuring it. Sometimes you can layer such a mixture putting the sand in first, the vermiculite or perlite next, and sphagnum moss on top; also put sand in the holes made, as suggested above. I happen to prefer mixing the ingredients. If you are working with small, fine stems or very small seeds, it is wise to put the whole mixture through a ½-inch mesh screen, especially if you are using peat moss instead of sphagnum moss, which is sometimes hard to get.

Make the same sort of sterile mixture for starting seeds indoors, though of course the value of the sand for seeds is not scratchiness but good drainage.

A starting mixture for the seeds or cuttings of acid-loving plants is often made from 3 parts of peat moss and 2 parts of sand. When you use a mixture like this, I'd recommend sterilizing the sand, for acid-loving plants are especially susceptible to pathogens, especially strange fungi to which they are not accustomed. (Acid peat moss is already sterile.) In the old days, leaf mold was often used for acid-loving plants, in combination with loam, peat moss, and a little sand. Then clean, sterile sand was also sprinkled on top of the flat after the cuttings were put in, to keep the surface sterile. Now vermiculite is often used instead, though sphagnum moss is also excellent because of its special qualities of acid and antibiotic sterility. (To make a supply of pure leaf mold for this mixture, rake oak, beech, and other leaves into a heap inside a board frame, with good spaces between the boards to provide air. Water and tamp down the leaves as you add layers. Do not add weeds, green stuff, or twigs, which will invite fungi to come in instead of the bacteria you want to work on the leaves. Also never add manure, nitrogen meals, lime, or any leaves taken from tarred roads. Ripen the leaves for two years and then use in all mixtures calling for leaf mold. You can also substitute leaf mold for peat moss in all combinations except those calling specifically for acid peat.)

Once started, rooted cuttings and seedlings grow fast, and nutrients are soon needed for the roots to draw on. Unless you have provided in some way for nutrients to be added to the starting mixture, you will need to move the young plants to a regular potting soil with loam in it.

One method for including soil in the starting mixture is to fill your flat first with a layer of broken crockery or pebbles and gravel. Then spread some nylon stockings over this to

*Potting Soils for Rooted
Cuttings and Seedlings*

hold up the next layer of soil, on top of which you put a starting mixture for rooting or germinating. With this arrangement you still have fairly sterile conditions in the area where the seeds or cuttings are getting established, but you also provide nutritive material the roots can reach down to as soon as they

Here a flat is prepared for rooting cuttings or starting seeds. (1) The bottom of the flat is lined with a layer of broken clay pots for good drainage. (2) Over this goes a layer of rough, rich compost mixed with soil. (3) And finally a layer of finely textured starting mixture.

3

start growing. These will be new, tender roots, so you must be sure that the soil is loose enough to let them through. You don't want to start their life in an environment where they have to struggle and be stunted because they cannot get through to what they need. Peat or compost added to the soil layer will help to encourage the roots to grow. Keep them encouraged, and give light additions of fish emulsion or seaweed solution or whatever mild organic fertilizer you wish. Setbacks at this early stage from lack of nutrients, water, and air can be very serious.

For best results, sometimes it is wise to make up mixtures of a wide selection of ingredients which can provide for several situations. These blends can very well include manure for extra nitrogen. One gardener I know keeps on hand a mixture of 2-2-2-1-1-½-½-¼-¼ of sand, soil, damp peat moss, vermiculite, dried cow or sheep manure, rock phosphate, cottonseed meal, potash rock, and ground limestone (unless she is potting for acid-loving plants). This strikes me as rather elaborate, but there is some logic in making such mixtures to be on the safe side.

Simpler is to have on hand some good manure tea, made by adding water to a mixture of 3 shovelfuls of horse or cow manure, 3 cupfuls of blood meal, and 3 quarts of compost steeped in a barrel of water (preferably rainwater). You keep adding the water until you have it diluted to the color of weak tea.

In England gardeners like to use finely crushed clay flowerpots as one ingredient of their propagating mixtures. A good British recipe for starting acid-loving plants is 4-4-4-2-1 of loam, sifted leaf mold, sharp sand, crushed pot, and peat moss. After the plants grow, or strike roots (as the English say), shift them to a mixture of 6-4-1-1-1 of neutral soil, leaf mold, peat moss, crushed pot, and sharp sand. The advantage of using the crushed pot, of course, is that it adds another ingredient which holds water well. Then, as the plants grow larger, enrich the mixture from time to time with bone meal, dried blood, and an old English favorite, hoof and horn meal. For lime-loving plants, English gardeners add mortar rubble or the equivalent in limestone chippings. Any of these ingredients in your store of supplies will give you the chance for more flexibility and more exact adaptation of your various mixtures to your needs than you will have if you stick to the old standard 1-1-1 of loam, peat, and sand.

The garden loam you might scoop up to put into a potting mixture may cause trouble, and is not going to be nearly so successful as pasteurized soil. It is imperative to know the difference between pasteurized and sterilized soil, and unfortunately people who give out gardening advice often recommend sterilizing. I say: do not do it.

Sterilization is usually at a temperature of 250°F. and that is high enough to kill all the beneficial organisms you definitely want to have in the soil to help your plants. It is much better to pasteurize, for *pasteurization* means heating no higher than 180°F. It is useful to know that when you steam soil at 120°F., it will be cleansed of nematodes; and that at 140°F. most plant-pathogenic worms, slugs, and centipedes will go; but that it takes 160 Fahrenheit degrees of heat to rid the soil of all pathogenic bacteria. At that heat the beneficial bacteria, it is believed, are not harmed, even after the recommended 30 minutes of steaming. Maybe 15 minutes of steaming is enough if you are only working with small amounts. And if you use a pressure cooker at 5 pounds pressure, you can reduce the time to 10 minutes.

The thing to remember is that too much heat is worse than none, for then you run the danger of killing all the living inhabitants needed for antibiotic and biochemical actions, and also of burning the organic matter, making it nearly useless for good soil health and good soil structure. In fact, if you use too much heat on peat moss or sphagnum moss, you might injure the moss and even create substances that can kill beneficial soil organisms. This could ruin the value of the humus.

Pasteurize the soil, therefore, *before* you add either kind of moss or compost. The compost or compost tea you add is likely to be free of pathogens. That's because the heat of a well-made compost heap will disinfect the materials, for the

Pasteurizing Loam

temperature in such a heap goes up to 145°F. or so. And remember that sphagnum moss is already sterile, and so are vermiculite (which is expanded mica) and perlite (exploded volcanic ash). Potting mixtures you buy at the store are also sterile, but I wonder what degree of heat has been used in the process. The sand you use, of course, can be heated to any degree you wish because it is inert and no harm will be done.

If you have already tried baking soil to get out the pathogens, you probably have noticed what an awful, unpleasant smell there is in the house while it is in the oven. One reason for this is that you undoubtedly had the oven too hot. At 160° to 180°F., the smell would be much less, and the results would be much better.

Testing for Humus

If you are in doubt about the organic matter or humus in the soil you plan to use in your potting mixture, put some soil in a bottle of water and let it settle. The minerals will go to the bottom, and if there is no perceptible layer of lighter weight organic particles on top of the mineral layer, you know that you will want to add some good compost, pure leaf mold, peat, sphagnum moss, or whatever is needed to amend the mixture. (This method can be used for testing soil for many uses.)

An alternative to pasteurizing soil in the oven is to do it in a double boiler on the stove, or, as here, on a hot plate. Fill a good-sized pot halfway with water. Then select a clay pot that just fits inside the pot and plug up the drainage hole (1). Fill the pot with soil, put it in the pot and heat it until a meat or candy thermometer inserted deep into the middle of the soil reads 170°F. (2).

A propagator who plans to make a lot of cuttings or to plant a lot of seeds indoors (or in a greenhouse) would do well to keep a supply of top-quality loam on hand, ready to steam for use. Keep the supply fruitful and productive by adding plenty of compost and nutrients as needed: examples are bone meal, dolomitic limestone, greensand, granite dust, and rock phosphate. Keep improving the mixture over the years. You will then have a supply of the best sort of medium when you add it to the sand, moss, and vermiculite needed for propagation.

The benefits of a good loam in a good mixture include a more even temperature in the mixture, an increased capacity to hold the soil solution and its nutrients, a moderated pH that is not too acid or too alkaline, a better habitat for the soil microorganisms you need, and a closer affinity between the tiny new roots and root hairs and the soil particles and nutrients made available by the microorganisms that cluster in great numbers around new young roots. Loam also provides enough air to promote the rapid oxidation characteristic of the very early growth of your new plant. This good soil, then, will have approximately 25 percent air, 25 percent water, 5 to 15 percent organic matter, and 35 to 45 percent mineral matter in it.

The best way to find out about the contents of your soil is to send it away to your state university in a container you can get from your county agent. The report you get back will tell you about the nitrogen, phosphorus, potassium, and humus content. You'll also discover whether or not your soil is too alkaline or too acid to be suitable for plants needing a more or less neutral medium to grow in.

If it is too alkaline (too limey), you can bring down the pH to a more neutral condition by adding materials such as seaweed, sawdust, pine needles, oak leaves, and leaf mold from acid-loving trees. You usually compensate for these by adding nitrogen-rich materials such as cottonseed meal, alfalfa meal, manure, or anything else suitable that is rich in nitrogen. You can also have on hand some Epsom salts to make a mixture of ½ teaspoon of Epsom salts to 1 gallon of water. When you add this to alkaline soil the water-soluble hardening factor, calcium carbonate, will be changed to gypsum, an insoluble substance that is harmless to plants. You can also add Epsom salts to your water if it is too hard to be suitable for delicate, acid-loving plants.

Send some of your pasteurized soil to be tested, too, to find out whether the phosphorus content has been reduced by the heating. If so, add rock phosphate, at the rate of 1 ounce per bushel.

Further details about starting mixtures and potting soils are given in chapters 2 and 6 on cuttings and seeds. Make up your favorite mixtures and then you can keep them on hand in boxes or bags.

The Right Environment for Propagation

Water, Oxygen, and Carbon Dioxide

Water, both in the mixture and in the air, is essential. Seeds cannot swell and grow without a good supply. Cuttings and grafts cannot "take" without it, for they then wilt and shrivel. To keep the air and medium moist, shade is often needed.

In nature a few leaves, twigs, pebbles, or clods of earth are often enough to shade tiny seedlings, but in flats or out in the garden, artificial shade is frequently needed for both seedlings and cuttings. Indoors it is easy enough to move flats into the shade or pull the curtains closed. Outdoors you can stick in small branches of brush on the sunny side of the plants or put on hot caps or homemade caps of newspaper. You can also build frames of lath. For large collections, a lathhouse like those at nurseries is the best solution. (See later section on Special Equipment for Keeping Cuttings Moist.)

Whether you're using pots, flats, or the special propagation devices described below, you can put plastic or glass coverings over cuttings inserted in the rooting medium. If you use plastic, it should be polyethylene or polypropylene. Do not use cellulose acetate, nylon, polycarbonate, polystyrene, or Saran®. Polyethylene and polypropylene are permeable to oxygen; all the others are not. (Plexiglas® is permeable to oxygen also, but is rather expensive.)

But plants need carbon dioxide as well as oxygen: the need for a good carbon dioxide supply was demonstrated by experiments done with cucumbers. When a source of carbon dioxide was placed in their propagation case for a period each day, the cucumber fruits were ready to eat within six weeks after germination. Since some plastics may be permeable to oxygen but not to carbon dioxide (or vice versa), make sure you use one that will permit the flow of both gases. Polyethylene and polypropylene meet this requirement. (In situations where more carbon dioxide is needed, you can use a cylinder or put small quantities of dry ice in a small closed space where it cannot injure the plants. In labs they use cylinders of CO_2. Ambient air usually has about .03 percent carbon dioxide, but the level can be raised to 1 percent or even up to 5 percent by dry ice.)

A proper plastic covering will not need to be removed during the rooting period. An improper one will. And a glass-topped container will also need to have some air admitted if the inside of the propagating chamber gets hot and muggy.

Mist-sprays, as described below, are often the best method to use to keep cuttings well-provided with a moist environment. This is a good method to use when the particular plant you are propagating needs a good deal of light and you'd be unwise to use shade.

Helpful Microorganisms

Once the roots emerge, you need microorganisms to aid in the biochemical processes of supplying nutrients to the plants. Luckily you do not have to work at this because, unless you have killed the beneficial bacteria and fungi with overheating, they arrive anyway. In fact, there will be 3 to 50 times as many of these microorganisms near the roots as there are in other parts of the soil. Because the soil in the garden also includes nematodes and other such pests, you may need to include marigolds in your garden environment once you plant out your new seedlings or cuttings; or you can provide manure fertilizers, which nematodes are said to avoid. (See earlier section on potting soils for suggestions on using manure.)

Equipment

Although they don't make the loveliest-looking planters, tin cans and plastic bottles with their tops cut off are fine for holding rooting and germinating mediums for propagating plants. Poke holes in the bottom for drainage. Heat the pick to make poking holes in the plastic easier.

A budding knife is essential if you want to make a good graft.

Pots and Bricks

Things you probably already have on hand can be very useful for propagation. Look over your resources and decide what other materials you want to invest in. If you need more pots, try used milk cartons, plastic containers, styrofoam cups, and tin cans, all of which are easily adapted if you poke holes in them for drainage. (Use a hot ice pick for the plastic containers.) But remember that when the containers are impermeable, you have to moderate your watering. Clay pots are useful because their permeable sides are good both for taking up water and for evaporating it slowly and gently. Remember, it is very important both for cuttings and germinating seeds to keep a stream of oxygen going through the medium and enough moisture in the medium and the air above it.

To keep the air around the plants moist, you will need plastic bags to prop up over the cuttings and seed flats to reduce evaporation. You also need some small sheets of plastic to wrap around the moss you use in air layering (see chapter 3), and for wrapping small grafts (see chapter 5). You can make strips from bags you have on hand. And again, let me remind you not to use Saran® plastic or any other material which is impermeable to either oxygen or carbon dioxide.

Other obvious things to have include: string, rubber bands, paper bags, good sharp knives—especially a budding knife if you plan to do that kind of work—some pencils for labeling and to use as dibbles, and either bushel baskets or big bags, trowels, a good sharp spade, glass jars, paper envelopes, the usual saws and pruners, perhaps some old lumber and old window frames and panes of glass to use for making a cold frame, hotbed, or propagation case. If you have any old, cracked, or chipped clay flowerpots, they are excellent to break up to use at the bottom of pots and flats for drainage. Of course you can use pebbles and gravel instead if you compensate for their being less porous than clay.

Clay pots come in three shapes, and makers of plastic pots tend to produce the same. The ones called *standard* are fairly high; *azalea pots* are somewhat lower; and the lowest pots are called *bulb pans*. All three types come in various sizes. Nowadays, there is also a whole array of cardboard and plastic pots and flats you can buy; and many kinds of peat pots, often called Jiffy® pots, after that trade name. Then there are Jiffy-7s®, the compacted and netted 2-inch tablets which swell up, when you water them, into little pot-shaped containers of peat moss. These are very handy for planting a few seeds or inserting small cuttings. (Be sure to wring them out a little after soaking, so they won't smother your plant material with too much water.)

The peat pots called Fertl-cubes® have commercial fertilizer in them, but you can make your own by soaking untreated peat pots (and the peat you use in them) in compost tea.

Also have some bricks on hand to use for growing very tiny seeds and spores of ferns. Soak the bricks until they are saturated, and then bake them for 30 minutes at 325°F. When the sterilized bricks are cool, spread each with moist, shredded sphagnum moss. Then sprinkle the tiny seeds or spores on top. Keep the brick in a tray of water so the moisture will keep rising by capillary action. You can also use little planting sets especially made for tiny seeds. These have 10 or 12 sections and a fitted plastic lid, and are available now from many nurseries and seedsmen.

These suppliers can also sell you blocks of attached peat pots, which are very convenient. Some come with a plastic tray for the pots, and that makes watering easier. A pane of glass or a plastic polyethylene bag over the top of the pots will help to maintain the high humidity needed.

There are cardboard and plastic trays now on the market for use as flats. You may find them easier to handle than the somewhat larger 24-by-18-inch, old-fashioned wooden flats we all used for so many years. But you can make smaller wooden ones yourself, and they are certainly recommended for cuttings which you root in a mixture containing a good deal of sand. Wet sand is rather heavy to carry, and two small flats are less of an ordeal to move around than one big, heavy

Another piece of equipment that will come in handy if you do any top-grafting is a cleaving iron.

Some good potting containers: right rear, standard clay pot; middle, azalea pot; and left, bulb pan. Foreground right and left, peat pots; middle, Jiffy-7s®.

Flats, Hotbeds, and Cold Frames

one. Boxes for cuttings should be 4 inches deep, but they can be as small as 8-by-10 inches if you want to make them that size.

Larger flats made of styrofoam to use with a series of three hoops or wickets and a plastic cover are also now available. With good drainage material, they are satisfactory. I use one that is also equipped with a heating cable to put in the bottom, and find it very handy for rooting cuttings and for starting seeds which need bottom heat of 70°F. If you prefer not to use an electric cable, you can make an effective hotbed by putting in a lower layer of horse manure. This gives off a lot of heat, enough for both seeds and softwood cuttings.

I'd also like to stress the value of providing yourself with a cold frame. They are valuable for starting garden plants and hardening them to outdoor conditions. Many bulbs and tree seeds which you may want to propagate must undergo a cold period before their dormancy is broken and a cold frame is an excellent place to put pots of seeds or bulbs, hardwood cuttings, or whatever you are storing. For excellent directions on how to put together an economical and permanent cold frame, see pages 63-68 of Sam Ogden's *Step-by-Step to Organic Vegetable Growing* (Rodale Press, Emmaus, Pa., 1971). If you have no cold frame, a cool cellar will also do.

In a cold frame, or cool cellar, as indoors, you can use trays with moist pebbles on top of which you set clay or plastic pots. The moisture from the pebbles will add humidity to the air around your plants and cuttings (as most house plant enthusiasts know). Do not ever use copper or zinc for the tray or for anything to do with plants because some of the metal dissolves in the water and this is not a bit good for them.

Propagation Cases

Terrariums, aquariums, and—if you can get one—one of those marvelous, old-fashioned nineteenth-century Wardian cases can all serve as good glass devices for maintaining humidity. But germinating seeds and young seedlings kept in them will need more airing than the mature plants ordinarily kept in terrariums or other such glass structures. You can also rig up miniature greenhouses out of panes of glass or make whole rows of them to cover cuttings put down to root outdoors. Now that we have polyethylene, too, any of these structures can be imitated in plastic. You can make them yourself by propping up sheets or bags of plastic on any kind of stakes or wickets you find convenient. Or you can buy them in the form of clear plastic shoe boxes, bread boxes, cake covers, or a very handsome 3-foot-wide, plastic domed terrarium I once saw for miniature begonias. For more air, you can poke holes in such plastic structures.

Some people use plastic bags to enclose fruit blossoms during the pollination period to keep away bees before hand-pollination. Others just use small brown paper bags or heavy gabardine ones, which are good for wind-pollinated plants. Any of these can protect flowers from the wrong pollen when you are crossbreeding.

You will also need envelopes and vials for storing the seeds you save, or the pollen you plan to use in a later season. Vials for storing pollen are usually stoppered with nonabsorbent cotton and a glass stopper. If the stoppers stick, use carbonated water, which can seep in and get them loose. Some plant parts you want to store need moisture, so you can use vials as vases with water or moist sphagnum moss in them. Many people who save their own seeds of vegetables and flowers make their own envelopes of folded paper, leaving the ends open for all seeds which need warm, dry storage. When you save tree and shrub seeds, have plastic envelopes on hand with enough moss to insert to keep them moist. (See Part II for specific examples of those which need these kinds of storage.)

Bags, Envelopes, and Vials

It is possible to plant and grow things with considerable carelessness, but you may have disappointments. I guess all gardeners have their share of carelessness and disappointment, but at least as far as cleanliness goes, why not take some precautions?

Cleaning and Sterilizing Equipment

People who are both careless and unheeding of organic gardening methods just douse everything with poisons when trouble arises, or even merely in anticipation of trouble. But there are better ways.

You can avoid a good many failures with simple soap and water, especially boiled water. It is sensible to boil your utensils, too. Plant materials cannot stand the 212°F. heat you need for boiling, so soap and lukewarm water will be best for them. Use Dial® soap or any other soap which contains hexachlorophene. For surfaces that are especially hairy, waxy, or sticky, add a little detergent to help make the water wetter. Clorox®, Lysol®, and sulfur are also useful to make solutions for sterilizing bulbs, corms, cuttings, or other parts of plants that you are going to store over the winter.

Special Equipment for Keeping Cuttings Moist

This clever device is designed to keep hardwood cuttings shaded as they root in the garden. Slant the light-reflecting white movable roof so that it keeps the sun's direct rays from the cuttings. In winter you can cover the sunken box with glass or plastic to serve as a cold frame.

Because you leave some leaves on softwood and semi-hardwood cuttings to help stimulate root formation, they must not be allowed to wilt and die before the roots grow. Some arrangement for moisture is often needed besides the plastic covering you put on to keep the soil moisture circulating. Mist-sprays are good because they help transpiration from the leaves to go on even before roots have formed. But you do not have to have a big greenhouse like those of commercial propagators.

Mist-Houses   You can rig up an arrangement outdoors with a sunken box large enough to hold the number of cuttings you wish to start at one time. Over it you put a white hood or sloping roof, facing north so it will reflect light but not let the sun shine directly on the cuttings. In summer or other seasons of warm weather, bring in a hose with a mist-nozzle to provide the high humidity which is always needed by plants under stress. In winter you can put a pane of glass over the box, in imitation of a cold frame, and put some insulation around the base of the box to help protect the cuttings from freezing. To get even more light put this imitation cold frame near a white wall which will reflect light, too. Do not use plastic instead of glass for winter protection; it can be ripped off in bad storms, and the freezing air that will get in can ruin all your cuttings.

A small mist-house to use in moderate weather was described by Don Merry in the November 1974 issue of *Organic Gardening and Farming*. He made his propagation house a 3-by-3-foot structure with a slatted floor. Around this he put 4-inch-wide boards to make a tray, and then over the tray he built a 2-foot-high little house with a gabled roof. (Merry put hinges on one side to permit entry and occasional airing.) Then he covered the structure with transparent polyethylene plastic. When he put in the hose, he found that the best way to do it was to make a hole in the floor and bring the nozzle and end of the hose up through the hole. Merry fastened the nozzle to the lower part of the roof, so the spray would be well above the cuttings when he inserted them. He says the best medium to use in such a mist-house is vermiculite for the bottom two-thirds of the starting mixture, with pasteurized soil or sand on top. By careful planning this propagator managed to fit in a hundred cuttings, and he even started some in peat pots. He rooted roses, euonymus, geraniums, and chrysanthemums in this little mist-house. He also was able to start many kinds of shrub cuttings. The amount of water used in a mist-house like this is about 1½ gallons during a misting period lasting from sunup to sundown.

Another handy device for heightening moisture around your cuttings is a U-shaped screen, open to the sky, with a mist-nozzle on the hose you fasten to the top of the screen. This device is good to use where the weather makes it desirable to have air available at all times for the cuttings, and where enclosure would make too hot or too muggy an atmosphere to keep your cuttings healthy. Actually this and the previous structure are variations on a terrarium.

Wigwams   To keep rooted cuttings from wilting after they have been put into the soil, you can also cover them with wigwam-shaped structures. (See illustration.) These structures will gather moisture produced by transpiration from the leaves and evaporation from the soil. Then this moisture condenses and drops to the soil to be reabsorbed by the plant. The cycle goes on when the moisture again leaves the plant through transpiration and evaporation the next day. A bell jar

A wigwam, like the one shown here made from reinforced plastic film and four strips of scrap lumber, provides plenty of protection for young garden plants.

makes a simple wigwam structure, but both this and the netted polyethylene wigwam need to be shaded whenever the sun gets really too hot. Use a cloth or piece of wet burlap. When you see that condensation has slowed down, add water. Remove the glass jar or plastic covering when the plants are no longer subject to wilt.

I hope that someday someone will try out an idea I had one day. Why not, I wondered, rig up a misting arrangement by using a humidifier like the one I rent in the winter from the drugstore to send unmedicated mist into the air to help my sinuses? Then why not cover humidifier and plants with a big, transparent, polyethylene bag? I know my sister-in-law has a permanent humidifier in her house, and her plants and cuttings do very well.

Benefits of Misting Devices   Misting or maintaining a high humidity around cuttings will result in good, strong root systems. Also, in case of wilt, or near wilt, any of the devices I've described will help to restore the plants. The air needs to be cool around cuttings because after rooting begins, there is an increase of respiration and cambium activity. At this time the air around the cutting should be kept cooler than the rooting medium. This cooler air helps make sure that leaf growth is held back and that roots are encouraged to form. If there is enough water in the mist to form a coating on the leaves of the cuttings, the transpiration rate is lowered and the leaves cool by as much as 10° or 15°F. This lowered temperature also makes it possible to give the cuttings very bright light, which is another great advantage in helping cuttings to strike roots and grow.

Besides fostering strong roots and good growth, misting devices speed up root growth. Sometimes two or three weeks are enough to promote roots that are large and plentiful enough for transplanting.

Lights   Plants, seedling plants, rooted cuttings, or whatever you are trying to grow away from direct outdoor light will probably need booster lights of some sort. You can grow seedlings, for example, on a windowsill, but all too often they shoot up too fast and twist towards the light because there just isn't enough of it. In the vicinity of New York City the bright sun at noon outdoors in winter on the shortest day of the year will provide 5,500 to 7,000 foot candles of light. But at the same time in the same area you get light more or less equivalent to the lower range of that, say 5,600 candles, indoors at the glass of a south window. When the plant is indoors but 1 foot from this south window, the light drops way down to 500 to 900 foot candles; to 350 to 550 at 2 feet; and to 200 to 400 at 3 feet. On the east or west sides of the house you can't expect more than 400 to 750 at the glass, and the light drops way down to 100 to 180 foot candles when the plant is 3 feet from the pane. The same kind of light can be expected in a north window, except that if the plant is right next to the glass, it will probably get 300 to 700 foot candles instead of the 500 to 900 you get in the shade on the south side.

I go into all this to emphasize why some artificial lights are really a help for plants. Two things are important to know. One is that the nearer the lights are to little growing seedlings, the better. (But not so near as to burn them, of course.) The principle is the same as for growing seedlings in a window: plants which do not stretch for light are shorter, sturdier, and healthier. The other main thing to know is that plants do best with a full spectrum of light; add an incandescent bulb or two to the two or three fluorescent bulbs you rig up (or have rigged up for you in one of the flower-cart structures you can buy). It's best to provide 1 watt of incandescent light for every 3 watts of fluorescent light. And use a reflector, for you might as well not waste the light, but get as much as possible angled back down on your plants. There are also full-spectrum lights you can buy, Vitalites®, for example. And Gro-Lux® and other lights emphasize the red end of the spectrum, which is good for perennial and annual seedlings.

It is handy to have pulleys for the lights mounted in the reflector so that you can let them down when seedlings are small and keep lifting them up as the plants grow. But you can always elevate your flats or peat pots on bricks or blocks of wood (or old books if your flat is styrofoam like mine) in order to keep your little plants at the proper distance from the light.

Plants should not be under 24-hour light, so if you decide on 16 hours, say, and want some help in remembering, there are automatic timers which will switch the lights off and on for you. They run about $15.

Measuring Light Intensity   Use a light meter such as those used with cameras to find out what the intensity of light is on your plants. Small seedlings of annuals like petunias, snapdragons, and verbenas, for instance, need at least 1,000 foot candles or more for 12 to 14 hours. Mature French marigolds will do well with long, 18-hour days at 600 to 750 foot candles. If you give that amount and degree of light, they will bloom all winter for you. With less light, short-day plants like cockscomb and salvia will do fine with a 10- to 14-hour day.

Just as valuable are the ways you can encourage cuttings by artificial lights. Rose and euonymus cuttings, fruit trees like peaches, and even the more difficult hardwood materials will do all right if given 5,000 foot candles of light, along with the bottom heat and the humidity that is maintained under a pane of glass or a plastic cover, or by mist. Take cuttings of about 4 inches, with 2 nodes as explained in the next chapter. At 1,000 foot candles, properly spaced, you can also grow radishes to plump, edible size in 6 weeks. (Space 5 seedlings to a 4-inch pot, sunk in moist peat or moist sphagnum moss in the flat.)

Regulated lights are very useful indeed for rooted cuttings of kalanchoes, poinsettias, and other plants requiring short days to induce flowering. At 60°F. and 500 foot candles, such cuttings will root and do well. Give them only 9 hours of daylight for several weeks to bring on flowering. Never interrupt the dark period, even with car headlights or an uncurtained streetlight; it is actually the length of the dark period rather than the length of the light period which is important. (And keep the roots moist constantly and both cool and moist when the buds form or they will blast. Never let the heat rise above 65°F.)

Rooting Difficult Cuttings Under Lights   Hardwood or softwood cuttings from a rose used to be thought of as among

A plant propagator checking light intensity of plant shelf with a light meter. Note the foil in the back of the shelf which reflects sunlight coming in through the window back on the plants.

the most difficult to root, but they can now do very well if kept moist and warm and under lights. Cuttings of rose hardwood taken around the end of September can root in the moisture and heat of a well-lighted propagation box in about a month. And softwood cuttings are even more pleasurable to root. If taken from crisp stems on a cool day in summer, rose softwood cuttings may well root in 10 days. When you plant them out, keep the rooted cuttings well watered and well weeded so that they can grow strong without any setbacks.

Other difficult hardwoods such as the broad-leaved ever-greens also do very well under lights and mist, or under lights and the humid, enclosed, homemade plastic propagation devices.

Pests and How to Outsmart Them

A good part of a gardener's expertise depends on a knowledge of the ills that can befall his or her plants, and this applies to the propagator, too. Clean all plants before starting any propagating and keep on the watch for pests large and small, from the ubiquitous snails and slugs and mice to all the little insects and baneful bacteria and fungi that may arrive if you do not take precautions.

Leaf spot and leaf blight on the mountain laurel you are trying to establish indicate that you'd better move your cuttings or new plants from the shady place where you have been keeping them right out into the sun. But lacebugs on any of the rhododendrons or azaleas mean you should do the opposite and move the plants out of the sun and into the shade. Also clean the plants again well.

In fact, make a habit of keeping your plants clean. When powdery mildew gets onto your asters either before or after you divide them, it isn't going to do much harm, but clean the asters anyway. Aphids you see attacking can often be overcome by hosing (early in the morning is best). You may discover borers or cankers on your dogwoods; Japanese beetles on your Virginia creeper, wild grape, and hardtack; or euonymus scale on your bittersweet, spurges, and, of course, euonymus. If so, give the plants a good dose of fertilizer and expose them all to air. Pests like borers can also be snagged out with a bent wire pushed down and wiggled in the hole. Fire blight, which hits roses, fruit trees in the rose family, and shadbushes, is taken care of by cutting back any affected branch to a foot below where you see the trouble. Fire blight is easy to see, for the leaves all suddenly turn dark brown in late spring or early summer.

Keep apart plants which transmit rusts back and forth. These bad pairs include red cedar and apples, so if you are propagating either, avoid the other. Also separate azalea and hemlock; asters and two-needled or three-needled pines; goldenrod and those pines; flowering currant and white pine, or gooseberries and white pine (in our state of Vermont it is against the law to grow currants at all); blueberries and fir or spruce; hemlock and hydrangea or huckleberry; rhododendron and spruce. Before you go in for any such possible combination, ask your county agent what laws apply in your area, or what combinations are considered dangerous. Avoid propagating such pairs.

You may also unwittingly introduce pests when you bring divided plants or other plant parts either into the house or into the yard from another area. Violets might introduce scabby anthracnose to your yard. Azalea or hemlock might have black vine weevil. And a terrible thing to bring into your area would be the petal blight of azaleas because you could start its spread all through your state or region. Horrible to think of.

Plants indoors are subject to some special troubles, like those I discuss below.

Damping-Off—a Fungus Pest

Damping-off is just about the most serious threat to your little seedlings when they first come up; it can even destroy your plants before they emerge. Although the same pest attacks mature plants, it is most serious for little plants growing in a warm, moist atmosphere. To combat damping-off, never use poor, small, undeveloped distorted or broken seeds; for all such seeds invite fungal decay. Next, provide light, dryness, and good air as soon as your seeds germinate. (The change in conditions somewhat duplicates outdoor conditions where damping-off is rather rare.) You can also prevent this fungus by sterilizing your seeds in vinegar and water (1 tablespoon of vinegar per quart of water) and always keeping your seeds as clean as possible.

Provided your seeds are not old and weak, you can also use a 1 percent solution of Clorox® to disinfect them. (See Sterilizing Seeds in chapter 6 for details.) Strong, viable seeds may also be soaked in hot water. Heat the water to nearly 135°F. pour it over the seeds and let it cool to 120°F. Soak small seeds for 15 minutes and large seeds for up to 30 minutes. Another way of controlling fungus is rapidly to dip seeds in a small sieve into ½ cup of rubbing alcohol mixed with ½ cup of water.

Keep the workbench and containers clean, too, by using a 1 percent solution of Clorox®. Or you might want to try an equisetum solution. This can be prepared by combining 1 cup of equisetum weed—known as horsetail—with 1 cup of water in your blender. Then blend on medium speed for a minute or so. If you can't find a sandy place where this weed grows, write to the Bio-dynamics people at Threefold Farm in Spring Valley, New York 10977, for a ready-made solution. If you do have the weed itself, it might help to bury pieces of it in the soil. The planting medium can be treated with these too. For added protection, you certainly should have a top layer on your flat or pot which is sterile, like sphagnum moss, or one which is sterile and scratchy, like coarse sand.

One more precaution is worth knowing about: a well-filled and therefore well-aired flat is less likely to get damping-off than one that is half or two-thirds filled, and muggy. Good drainage is an essential. If none of these precautions stops the damping-off fungus, and you do see it arrive, you can use a tiny spoon, the end of a label stick, or even a toothpick to pick the fungus off.

Insect Pests

Healthy soil and healthy plants are not attractive to insect pests, but you still need to be on the alert for some invaders. Aphids disappear, as already suggested, if plants are well watered, kept moist, and well nourished both indoors and outdoors. As you may have heard, a good preventive in the garden is to introduce ladybugs, who will eat up dozens of aphids and cut the population back. When you buy ladybugs, ask for those entering the phase when they eat aphids, for if they're not in this stage of the life cycle, they may just fly away. Also, place them at the base of infected plants early in the morning or evening when they tend to be less active. The quantity to buy is 1 quart or so per acre. Sprays of fine clay or of the essence of nasturtium also have been helpful at times in controlling aphids to some gardeners. Fine-needled diatomaceous earth makes a good spray, but if none is available, you can substitute eggshells put through the blender. Some of the aphid-harboring plants you might weed out are chickweed and lamb's-quarters. Two that also attract aphids but that you probably will not want to get rid of are daphne and lettuce.

If you find that groundsel or wild white yarrow nearby is attracting brown lacewings to your propagating beds you'd better do something about it. I have never found brown lacewings on my cultivated yarrow, but it will pay to keep your eyes open if you have either. If necessary, get rid of the weeds or move the bed.

For nematodes there are two often-employed controls: repellent plants like asparagus and the especially effective Mexican marigold (*Tagetes minuta*) and the attacking fungus that is usually present in manures, rich composts, and mulches.

Cutworms can appear overnight to attack the small plants you put out in the nursery or vegetable garden. One age-old way to foil their habit of biting off your plants is to encircle each plant with a cardboard or tin-can collar. Another is to put a toothpick stake beside each little plant so that the worm is frustrated when it tries to encircle the stem. A cutworm is also likely to be repelled by sharp sand or an oak-leaf mulch put around each little plant. A hundred years ago people trapped cutworms by putting down handfuls of milkweed, clover, mullein, or elder sprouts to entice them to gather. Then they were collected and done in.

Other nasty little worms can be sidetracked by pieces of carrot or raw potato placed near the tender plants. And for bigger, even nastier earwigs, squash bugs, slugs, harlequin bugs, and snails, you can put down boards. The pests can be found under these in the early morning. (A smearing of molasses on the underside of boards will help lure them in.)

Or keep chickens. They are among the best bug controls there are. Or buy and introduce *Bacillus thuringiensis,* a bacterium which kills 110 species of pests.

Some of your cuttings may be infested by mealybugs and scale insects. You can wash the infested places with soap and water, which will eliminate some; then dip a cotton-tipped swab in alcohol and touch the pests. For an even more drastic treatment you can wash off the leaves with a kerosene emulsion, though this also will eliminate good insects like ladybugs if they happen to be there at the time.

Young cucurbit plants like squashes and cucumbers often get infestations of squash bugs and their nymphs at our place. We put out wood ashes right away, both on the plants and on the ground immediately around them. This really sends them away. And if you want to go further, you can use rotenone or add hydrated lime or a sprinkling of turpentine. (This may harm the leaves, so shake off the bugs before you squirt them.) To discourage cucumber beetles, sow radishes and nasturtiums right in the hill with the cucurbits.

At our place we need special vigilance to control slugs. The best way to use the beer trick to entice and drown them is to sink a sardine can with sharp top edges down in the earth so that the edges are just level with the surface. This method is much better than merely putting down saucers, which the slugs can slither out of before they succumb. When you see a slug soon after daylight or uncover one while working among your plants, simply drop it into a saucer of salt. The slug will die in a few minutes because the salt will draw out the creature's body juices by osmosis. (You may be surprised to see that the slug turns orange.)

We also take the precaution of planting strong-smelling herbs and annual flowers among tender plants that are in danger of being invaded. But each year that the soil is built up and enriched with organic matter the danger, of course, is lessened. Some years the only pests we have (except for slugs, moles, and mice) are the squash bugs, a few Japanese beetles, and in very dry weather, a sudden invasion of aphids which we promptly dispel with a hose.

For pest control and pleasure, we give a lot of attention to keeping birds at our place. (Part II entries include plants which lure birds to your yard, a fact sometimes mentioned in the descriptions.) Even so, I wish we had more owls. For then they might eat up the mice, which are real pests, especially in the way they devour the tulip bulbs I like to grow and propagate. I keep a mouse trap going in the house, and I suppose it would be sensible to put some traps around outdoors where I suspect they run, for example, up and down the mole channels.

I don't mind moles, for I know they are carnivorous, but if you like, you can plant large, smelly bulbs of crown imperial (*Fritillaria imperialis*) down deep to scare them off. Some people say moles stay 10 feet away in all directions, and the bonus is large, handsome, big-flowered plants to brighten up your spring garden. Castor bean plants are also repugnant to moles. And wire cages around susceptible bulbs are a good protection against both moles and mice.

Many a young plant, carefully grown from germination to succulent seedling, has succumbed to the woodchuck. All the gardeners I know in this part of the country try one thing after another to keep these hungry herbivores out of the garden. Fences do help, especially low electric fences, though many householders do not want such things in the yard. Of all the repellents I have tried, hot red pepper (not cayenne) has worked best. It does wash off and blow off, but when it's on the leaves of a young plant the woodchuck does not eat there. I have also discovered that where I permit tomato volunteers to grow up to create a barrier, the woodchucks do not go through to the young plants growing beyond. Or did not the summer that the volunteers were allowed to stay. This seems a logical preventive, for woodchucks will clean out the beans right next to tomato plants and never touch the tomatoes. (See chapter 6 for other suggestions about protecting seedlings.)

Chapter 2
Cuttings

Probably the most frequently used and least troublesome way of propagating plants is the vegetative, or asexual, method of taking cuttings. With this method, you wound and cut off certain portions of the plant, and then treat them in such a way that they will grow into a new plant by sending down roots to support the stem and leaves; you can also cut portions of roots, and they will send up shoots. Leaves can send out both shoots and roots. Where a new plant has already been started by the old one, your cut will separate new growth rather than stimulate it.

Usually cuttings are spoken of as softwood, semi-hardwood, and hardwood; softwood cuttings can be divided into the kinds made with the tip of the stem included and those that are made from pieces of stem. Cuttings are also named for the part of the larger plant structure they come from: stem, root, leaf, sucker, tuber, offset, or runner.

The plant material used in taking cuttings has usually just had some active, vigorous growth. This can be either natural growth in the greenhouse or garden, or forced growth induced by pruning back a plant to get it to send out new shoots (like the pinching you do to summer annuals to get them to bloom well). Good cutting material, of course, should be disease free. It should never have been girdled by rabbits, or defoliated, or injured by insects, freezing, or drought. Also, cutting material should never have been stunted by lack of nutrients or excessive fruiting.

The Principles of Plant Growth

Meristems, Cambiums, and Calluses

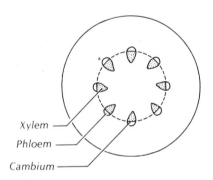

Xylem

Phloem

Cambium

Dicot stem with cambium layers arranged in a ring.

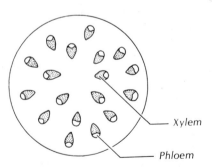

Xylem

Phloem

Monocot stem and its vascular bundles lacking cambium in a scattered arrangement.

To talk about how and where new cells grow, it's necessary to distinguish between flowering plants that have two seed leaves (dicotyledons) and those that have one (monocotyledons). Dicot plants have two major embryonic, or meristematic, areas where new cells are produced. The apical meristems are found at the shoot tips, or buds, and at the root tips. The lateral meristem—also called a *cambium*—is a cylinder within the stem. This vascular cambium is only one cell layer thick, but it produces two kinds of cells. Phloem cells are generated toward the outside of the cylinder; their function is to conduct food made by the leaves down the stem. Xylem cells are produced toward the inside; their function is to conduct water and mineral nutrients up the stem. Woody plants also have a layer called the *cork cambium*, which produces the corky cells of the outer bark.

On the other hand, monocot plants like lilies, grasses, orchids, and palms have no cambium. Because of this, monocots experience little or no growth in stem diameter, and they are understandably difficult to propagate through cuttings.

Like cambium, *callus* is extremely important in vegetative propagation. In fact, callus is derived from the cambium or from the adjoining young cells of conducting tissue. Forming at the injured surface of a plant, it is a tissue made of thin-walled cells. Botanists now know that callus must be formed before a cutting can root, and much of your success with cuttings will depend on how well you promote the growth of the callus.

Some plants send out roots very readily, and all you have to do is insert their cuttings in the rooting medium. (A slightly acid pH of 6 is favored for callus growth.) But other cuttings need quite special treatment and may need to be left out on a table for a day or so, exposed to the air, in order to form a callus. This technique is especially necessary for plants which have milky juices or are quite succulent, such as geraniums (*Pelargonium* spp.).

In plants like English ivy, peperomia, sweet potato, and the sedums, roots grow directly from the callus tissue, but in most cases they do not. Rather, the shock of being cut from the parent plant causes some of the cells of the stem to form a new meristem similar to the apical meristem of the root tip. The cells produced differentiate into all parts of the new root, including phloem and xylem cells which connect with the vascular tissues of the stem. This cell differentiation allows water, mineral nutrients, and sugars to travel up and down. People who have watched the process under a microscope have seen roots emerge in three days on chrysanthemum cuttings and in five days on carnation cuttings. It takes about 10 days, however, for the new root growth on chrysanthemums to be visible to the eye and nearly three weeks for carnations.

Some Variations

On plants like willows, poplars, currants, and hydrangeas, there are tiny knobs called *root initials* where roots start very easily. In fact, if you peel some bark off the stem of a willow, you can see small bumps. These are the spots where roots will grow, as many have found to their surprise when they only meant to use the stems for fenceposts.

Leaves of kalanchoes will develop new roots and leaves when leaf cuttings are made, with the new plants coming from latent primary meristems on the edge of the leaves. African violets and rex begonias can grow new plants from older mature cells at the base of the leaf or leaf stalk (petiole), but the roots must grow for several weeks before the new shoots begin to come. To insure that both roots and shoots will grow on jade plants (*Crassula argentea*) and rubber plants (*Ficus elastica*), propagators make a heel cutting which includes a bit of old stem and an axillary bud where the leaf stalk meets the stem. Some shrubs require this precaution, too. (See section on Hardwood Cuttings.)

Growth Regulators

All aspects of plant growth and development are influenced by plant growth regulators, or plant hormones. Common growth-promoters in this group are auxins and gibberellins. They are prevented from being overactive by inhibitors.

Auxins are important in the rooting of cuttings. Since an auxin is produced in the stem tip and is normally transported in a downward direction, it accumulates in the cut end of the stem and stimulates the growth of the roots forming there. In practice, cuttings are in one of three conditions when they are taken from the parent plant. They may have plenty of auxin action themselves and enough deactivation of the growth inhibitors so that new roots will easily form after wounding. Or they may have some auxins but not enough, so that an added natural auxin will help. Or they may be so lacking in enzymes to help synthesize roots that no added auxin will help.

Indoleacetic acid (IAA) is a natural auxin, and some propagators encourage root growth by using a powdery dip of indoleacetic acid and talc on their cuttings. Laboratory people prefer the synthetic auxin naphthaleneacetic acid (NAA) because it is not affected by the inhibiting enzymes in the cutting that destroy some of the natural auxins. Chemists have found a second synthetic substitute, indolebutyric acid, which together with talc goes into the root-promoting materials sold in garden centers. But if you prefer natural substances in your house and garden, you can probably obtain indoleacetic or indole-3-acetic acid through a science-supply house.

The growth-promoting *giberellins* are named for a toxin secreted by the fungus *Gibberella fujikuroi*. Causing a characteristic excessive growth in rice plants, this toxin was first isolated as a crystalline active material. Later it was regarded as identical or very similar to a substance related to flowering in the higher plants. Gibberellins are thought to be especially active in the formation of the long stems of flowers. (They will even convert the famous short peas of Mendel's experiments to tall peas.)

Since other gibberellin-caused increases have been observed, botanists think that its role is to remove a previous limitation or inhibition. For example, if the cold stimulus effective in producing flowers on rhododendrons is replaced by a gibberellin stimulus, flowers will come from that, too.

Auxins and gibberellins are essential to many kinds of plant growth, but growth inhibitors are a necessary counterbalance to keep your plants from filling up a room or yard, as they might unless they are controlled. In fall these inhibitors gain ascendancy over the growth-promoters and cause dormancy, so the plants can rest. This is the state you want for your hardwood cuttings. When the inhibitors decrease in activity in spring, bud dormancy weakens. Then new growth can begin with the ascendancy of the auxins and gibberellins. The breaking of dormancy is fairly complex, but many of us have done it in a practical way (with forsythia, for example) by bringing branches into the house in mid-winter, soaking them in the tub for a day or so, and then putting them into vases in a warm place to force blooming. Of course, this treatment doesn't work with some shrubs because their flowering is also controlled by such factors as day length and the need for alternating hot and cold spells to break dormancy.

Proliferation in Odd Places   Occasionally a plant wants to proliferate new plants in places where most others of its species never do. This happened with daylily stems given me a year ago by my garden club friend, Clare Sheppard. Perhaps it was an anomaly in the stem tissue, but embryonic or plantlet activity occurred halfway up the flower stalk. This development turned out to provide a perfectly good way to propagate these two varieties ('Pappy Gates' and 'Pink Perfection'). When I brought the stems home and put them in a light, sandy potting mixture, they took hold easily. It was only a matter of a week or so before I was able to cut away the old stem and have two new plants, or rather plantlets. (The usual way of propagating daylilies is by simple division. See that section in chapter 4.)

Take stem cuttings from those parts of the plant that have been in full sun, so that photosynthesis has provided the stem

Stem Cuttings

Choosing the Best
Stem Cuttings

with the maximum amount of sugars and starches. This stored carbohydrate food is necessary because the formation of callus and roots requires plenty of energy. For instance, a study of hydrangea cuttings showed that, protein synthesis was rapid and vigorous, increasing by up to 100 percent during the first 4 or 5 days. Now *that* takes a lot of food energy!

You can observe for yourself the significant effect of biochemical composition on plants to be rooted. Just take stem cuttings from two kinds of tomato plants, for example. A cutting from a leggy, pale seedling that is low in nitrogen compounds but high in carbohydrates will develop good roots but poor shoots. Conversely, a cutting from a stockier, greener plant will give good shoots but fewer, less sturdy roots. You can also demonstrate this point with chrysanthemum cuttings, for those with the highest storage of carbohydrates in the stems will root most easily.

Softwood Stems Should Snap According to old-time advice about taking softwood cuttings, the condition to look for is a stem which snaps easily when broken: this is really evidence of high carbohydrate content in the stem. If the condition is not right for rooting, the stem merely bends over and does not break. This means it is too young and has too many nitrogen compounds, that is, more than will be needed for the protein synthesis accompanying root formation. With all cuttings, it is best to withhold nitrogen fertilizer and to see that the plants are in full sun, making plenty of sugars. The best shoots are likely to be the laterals rather than the main stems, which are still growing rapidly and not just storing starch.

Use Young Plant Material For cuttings the younger the plant is, the better. One-year-old plants are better than 2-year-olds, and juvenile wood from trees and even from seedlings has been used successfully. Juvenile growth can be kept on older plants by regularly shearing them, as you would a hedge, to prevent the shift to the mature phase. Juvenile stems are perhaps preferred for cutting because inhibitors

30

may increase as the plant ages, or perhaps because chemicals that act with auxins are at an advantageous strength in younger plants. One good way to provide yourself with juvenile tissues is to make root cuttings first, then use the adventitious shoots growing up from them for your softwood stem cuttings. Remember, though, that tissues of young shoots propagated from adults are not true juveniles, for they do not regain juvenile characteristics.

Other examples of characteristics carried over from parent plant to rooted cutting include the habit of uprightness. Take, for instance, an upright Japanese yew (*Taxus cuspidata capitata*). If upright stems on this plant are used for cuttings, the resulting tree will be upright, but if horizontal branches are used, the resulting plant will not be upright. The same sort of thing will happen if you make a cutting from droopy branches on a coffee bush: the resulting new bush will be prostrate and spreading.

The Stem Section that Roots Best   Sometimes, especially with hardwood cuttings made from more mature wood, propagators cut the severed branch into several pieces in hopes of getting a new plant from each. It has been observed that with some hardwood cuttings the lower part of the stem, which is farther away from the tip leaves, is the section that roots best. This may be because there are fewer and fewer root initials as you approach the growing tip of the stem. When this basal section is best, it may also be that the necessary carbohydrates—as well as auxins or root-promoting substances—have accumulated in that part of the plant. When the tip roots best, as with softwood cuttings, it is probably because more cells in that part of the branch are meristematic and thus capable of new growth.

Usually Avoid Flower Bud Stems   Usually a cutting will not send down roots if there is no bud on it. This is because auxins are produced in the apical meristems, and, of course, buds contain apical meristem cells. But make sure you choose shoots *with leaf buds only* for cuttings. In other words, you should usually avoid the stems with flower buds. Re-

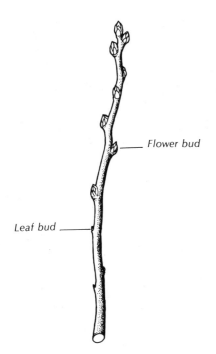

Flower bud

Leaf bud

A blueberry twig showing the difference between leaf buds and flower buds. The latter are not used for rooting stem cuttings.

Softwood Cuttings

moval of the flower buds will be of no help; the cause lies deep in the nature of the growth pattern of the stem.

With tomatoes, however, it makes no difference which kind of shoot you pick for a cutting. In fact many gardeners who prune back their tomatoes use all the clippings to start new plants for a later crop. And pruners of hedges also use their clippings—whether or not they have flowers—for cuttings to start new plants. Some of the cuttings will root, and when they do, the vigorous new growth the next year will make splendid material for new cuttings.

Don't Take Softwood Cuttings from Dormant Plants Avoid taking cuttings of softwood when the plants are dormant; often the best time is spring when there is sturdy active growth, or early fall after the summer's new shoots have hardened up somewhat. This stands to reason, since the inhibitors "control" a dormant plant, and the growth-promoting substances like auxins are more highly concentrated in actively growing plants.

There are some exceptions, though, to the rule of taking softwood cuttings during active growth. In the fall, for instance, you can pinch off the top growth of neglected petunias to start cuttings. Marigolds can also be taken in as cuttings in fall, then potted up for the winter as house plants. They root very easily just in water.

The conditions for rooting softwood cuttings are known to all practical gardeners: they are warmth, humidity, light, and a stem that has food enough to sustain the cutting during the root-forming period. When I said that a desirable stem should snap, I meant that it should neither be so young and flabby that it lacks strength and a food supply, nor so old and fibrous that it has exchanged food tissues for woody ones.

The length of a softwood cutting depends on the plant, but it is usually 2 to 5 or 6 inches. Sometimes the very short ones are called *slips,* as though these are sections of plant you just slip off as you walk by; which is, I guess, what quite a few people have been known to do in others' gardens when they didn't have the courage to ask for a cutting. But if you are tempted, do not just tear off a length of stem. Always make a careful slice with a very sharp knife so as to hurt the plant and the cutting as little as possible. If you can, take the cutting from the end of a nonflowering shoot, somewhat on the diagonal, and cut a short distance below a node (the point at which a leaf is attached to the stem). Sometimes you will discover that a plant roots well if cut about halfway between the nodes, but usually a cut below the node is better. Remove the leaves on the lower two-thirds of the cutting with a razor held rigid for a quick, sharp cut.

If you take cuttings when away from home, plunge them into cold water. If that is inconvenient, you can wrap them in wet papers for as long as 30 minutes to an hour. This does not apply to geraniums and other stems with milky juices, which should be exposed to the air for 24 to 36 hours to encourage callus to form.

Though garden perennials provide especially good material for cuttings in spring when growth is vigorous and healthy, you may take cuttings any time you can find young, nonflowering shoots on the plants. Shrubs that can be used for softwood cuttings in the spring after blossoming include lilac (*Syringa vulgaris* and var.). Among those that will root by September if started in the middle of June are winter hazel (*Corylopsis pauciflora*), broom (*Cytisus praecox*), Japanese

Cuttings are taken here from the softwooded polka dot plant (1). Take a good-sized cutting if you can, not less than 2 nor more than 6 inches is best, and make a sharp, clean cut. Then remove the bottom leaves, being careful not to leave any ragged edges (2), and place each cutting in the rooting medium (3). Sand, vermiculite (shown here), and peat moss are all good. To keep the humidity high around the plants, insert a stick in the rooting medium in the center of the cuttings (4), and bring a plastic bag up over the pot and its contents. Fasten the top edges of the bag around the stick (5). Occasionally open the bag to allow a little fresh air to circulate around the cuttings.

andromeda (*Pieris japonica*), and various azaleas, box, and viburnums. (Viburnums may be started as late as July.)

When you have your softwood cuttings in hand, keep in mind that though some stems will root in plain water, the usual rooting medium should be as coarse and capable of holding air and moisture as sand, vermiculite, and peat moss. (See page 2 for recommended proportions for ingredients. Use clay pots, peat pots, or double pots. See illustration on page 11.) And avoid components like unsterilized soil that might promote fungus disease. Other precautions against disease are equally important when handling cuttings. If you are unsure about the cleanliness of the plant material, dip it in a weak solution of Clorox® (1 part Clorox® to 5 parts water) · for 5 minutes. Use a milder solution for very young, sensitive cuttings made from seedlings. Detergent and water sprayed on the plants before making the cuttings can be helpful, too. If all goes well, your softwood cuttings will take from 2 to 6 weeks to root, less time than is needed for semi-hardwood and hardwood cuttings.

Semi-Hardwood Cuttings

Semi-hardwood cuttings come from wood that is older than softwood but not quite ripe. Plants used for such cuttings include hydrangeas, roses, pyracantha, ceanothus, cotoneasters, and some broad-leaved evergreens such as evergreen azaleas and gaultherias. All these and other evergreens such as arborvitae, juniper, hemlock, and so on can be taken for cuttings as late as October, November, or December.

Other shrub cuttings that should be taken late in the season for greater success include those from Oregon holly grape (*Mahonia aquifolium*) and many roses. (*Mahonia*, by the way, is one plant which succeeds better if the cutting is taken between the nodes.) Unless you have equipment for continuous misting in the daytime there is not much point in trying cuttings of magnolia and rhododendron. The preferred method for rhododendrons is layering. See Simple Layering in chapter 3.)

Your semi-hardwood cuttings will do best if you take a piece of short lateral growth with a heel, that is, a bit of the larger stem from which the small stem-cutting is taken.

It is customary to use the tops of the shoots selected for making the cuttings, but there are advantages in using the basal sections, also. (See earlier discussion of The Stem Section that Roots Best.) Cut the stems early in the morning when they are turgid, and cut just below a node. Keep the cuttings moist. Water loss is a big problem with semi-hardwood cuttings, so a mist-spray arrangement is almost obligatory. Bottom heat, as explained below, is often helpful to encourage callusing and rooting.

For best results with semi-hardwood cuttings take a piece with a heel (a bit of the larger stem) with the cutting, as has been done here with this rubber plant cutting (1). Then insert the cuttings in a good rooting medium (2). These large leaves need support, so they are wrapped around a plastic straw. When all are in place, the cuttings are slipped in a plastic bag held up with wire supports (3).

Because semi-hardwood cuttings take a long time to root and sit a long time in the cold frame or some similar place where they can be protected from too much exposure over the winter, many people do use indoleacetic acid in talc. You can dip the stems lightly in the powder before inserting them well down into the sandy rooting medium with only the tips showing. As soon as they are well fitted in the flat, a good soaking will help to settle them in and get the stems of the semi-hardwood cuttings into close contact with the sand.

Storage in a largish polyethylene bag is satisfactory. Be sure not to let the cuttings freeze, and if the cold frame is precarious, move them to a shed or cellar where the temperature will not go much below 40°F. You can use a mulch of hay to keep the cuttings warm.

Some needled evergreens are quite hard to root. These include spruces, firs, pines, hemlocks, and upright junipers. Therefore it is best to use young seedlings, and to take the cuttings in the late fall or late winter, and to work fast and put the cuttings at once where they will get plenty of north light and plenty of humidity. Remove the lower needles and apply bottom heat for a temperature of 65° to 80°F. Misting is essential for success with most needled evergreen cuttings.

Yew cuttings do better if either a heel or a mallet (whole piece of the larger stem) is left on and an additional slit or wound is made. (See section on Hardwood Cuttings for specific techniques.) The small stem you cut need be only about 3 inches long, though sometimes you can't find an appropriate one to use that is less than 6 inches long.

Semi-hardwood cuttings of broad-leaved evergreens are also made with some leaves left on. Use partially matured wood of such plants as azaleas, gardenias, holly, euonymus, and pittosporum and insert the cuttings into 3 to 6 inches of the rooting medium. Half perlite and half peat moss is a good one to try for these.

Hardwood Cuttings

Cuttings of hardwood are made on a very different principle, for you usually take them from leafless stems of dormant wood of the previous year's growth, though it's all right to use 2 years' growth for olive and fig tree cuttings. (For other exceptions see fruit entries in Part II.) Here you do not rely on the low level of photosynthesis to keep the stems going. Instead you make the most of dormancy, just keeping the cuttings alive and presumably callusing until spring.

As with other cuttings, be sure to choose disease-free plants which have never been injured by frost, hard freezing, drought, and insect defoliation and never stunted for any other reason. Use branches that have grown in full sunlight, unshaded by other trees or shrubs or by other branches on the same plant. And do not use rank growth with long internodes or small, weak shoots. All pieces chosen for hardwood cuttings should have a good supply of stored food to carry them over for the winter. Since tip sections usually do not have that good supply, concentrate on using the basal and middle sections of the stems. If a cutting is to be used later as a rootstock for grafting, be sure to put the scion onto the original cutting, not onto any new shoot that may come from one of the buds. The most successful hardwood cuttings are from ¼- to 1-inch in diameter, although 2-inch cuttings can succeed if no other sizes are available. Take cuttings in the early morning while the stems are still moist and full of water.

Cut the good, healthy stems you are going to use into convenient lengths, such as 7 to 10 inches. Each stick should have about 4 buds, and the cuts of the stem should be made just above a bud at the top and just below a bud at the bottom.

Sometimes instead of taking straight sticks it pays to take hardwood cuttings with a heel or a mallet. The heel consists of a short section of bark from the main stem which is attached to the small shoot that is removed for the cutting. The mallet is a whole piece of the larger stem attached to the 3-budded or 4-budded small stem. (Cuttings taken with a heel or mallet often have a terminal bud. If the plant is a hibiscus or weigela, the cutting certainly should include the tip bud, for those plants, among others, rely on the growth substances from the bud to help them strike roots.)

Bundle the cuttings with the tops facing in the same direction. As with other cuttings, it is best to put hardwood cuttings in close proximity; their hormone, or antibiotic, or nutritional, or other exchanges—whatever they are, and no one quite knows—have been observed to benefit the plants. The fact that you're using harder, older wood of one year for such cuttings provides another advantage: they can go deep into the ground and hibernate in safety, never getting too dry or too wet unless there are floods or torrents.

Since you're aiming for a good callus by the spring after you take a hardwood cutting, it sometimes pays to encourage callusing. You can do this by burying the pieces horizontally, or even upside down, in the medium where they are to spend the winter. For such cuttings this should be very sandy soil, pure sand, or pure sawdust (preferably somewhat aged). Callus forms well when the medium is slightly acid, so you need not fear sawdust. Burying upside down is probably helpful because the basal end is within a few inches of the surface of the medium, and thus gets more air and perhaps more warmth.

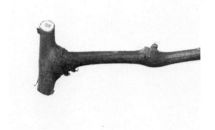

A heel cutting (1) is the shoot you wish to root and a small piece of the main stem. A mallet cutting (2) consists of a shoot also, but instead of just a piece of the main stem bark, a whole piece of the stem is left attached.

To root hardwood cuttings, (1) cut dormant stems just above a node, and (2) prepare 7- to 10-inch sticks. (3) Gather them together so that their tops all face the same direction, and (4) tie them together and lay them in the bottom of a flat. (5) Cover them with rooting mixture, (6) making sure the medium gets into all the crevices. (7) Then dampen the buried cuttings.

If the cold frame you have can be kept at 45°F., hardwood cuttings can be wintered there, but often a cellar or other cold room is a more convenient place to put the cuttings, whether they be in a flat or in a plastic bag. During the storage period the bottom end will usually callus. It may take two years. That callus will promote root formation when you plant the cuttings out in the ground in spring to come to life again and start proliferating new growth. The planting out is done with only the tip showing above the soil line. If, however, you have more than four buds—as sometimes happens with long shoots of apple—you can let two or three buds remain above-ground to form shoots. (See also the section on Putting Cuttings Outdoors.)

Like many softwood cuttings, a few hardwood cuttings can be put in water instead of sand or a rooting mixture. You can do this with easily rooted plants like willows (*Salix discolor* or *S. tristis,* for instance) and small boxwood (*Buxus sempervirens rotundifolia*). Depending on the length of the cutting, put them into 1 to 3 inches of water. They can be left in the water and watched until they form roots. When a lot of roots do form, it is often rather messy to get them out of the water and into a potting medium without having them get all tangled. It is a good idea, therefore, to add small amounts of sand and vermiculite, for example, to the water over a period of a week or so. This will give the roots something to cling to before being moved and also help them make the transition to getting oxygen from air in the soil instead of from water.

Mounding and Girdling   You might want to give a boost to some stems you want to use for cuttings by trying a very interesting practice that has been used for over 100 years. It involves mounding dirt up around the base of stems to keep them in the dark in order to get them to root. The same effect can be created by trenching in a way similar to one kind of layering. (See Trench Layering in the next chapter.) Though the darkness reduces the carbohydrate content and thickens the cell walls of the stem, it does make the stem more responsive to the indoleacetic acid sent to that part of the plant. Old mature ivy tissues, usually very difficult to root, will do so in the dark or in very low light. This is probably because auxins are destroyed or put out of action by light.

Instead of piling up soil, sometimes propagators put tape on the stems in hopes of achieving the same result. Or they girdle the stem or constrict it with a wire in order to block the downward passage of sugars and starches and concentrate them in the area from which they want roots to grow. With some citrus plants the practice is to girdle or bind the stem several weeks before taking the cuttings.

Using Boosters   Of all the different kinds of plants you may use for cuttings there are a few which will benefit more than

Helpful Practices for Stem Cuttings

the others by a dose of a growth-promoting substance. These include azaleas, blueberries, hollies, flowering dogwoods, yews, roses, cotoneasters, daphne, and magnolias. (Those probably not helped include boxwood, begonias, chrysanthemums, and zonal geraniums.) Plants like euonymous, fuchsia, boxwood, ivy, and lantana benefit from a dip in a high-analysis fertilizer, such as fortified compost with rock phosphate and soy or alfalfa meal added. If desired this can be in liquid form. Bottom heat of 65°F. is also sometimes helpful.

Preventing Early Budding Whether you take cuttings to root in summer or over the winter, one thing to watch out for is developing buds. Unless you keep your hardwood and semi-hardwood cuttings at a constant temperature of 40°F., they may begin to send out new shoots. Do not allow this to occur, because cuttings will die if they develop shoots before they have a chance to send down roots. You may not see the consequences right away, but after planting out the cuttings in the soil you will see that they will die. (When you do put them out, see that the cuttings are not molested by rodents or by weeds.)

Applying Bottom Heat There is one exception to the rule of maintaining a 40°F. temperature to prevent early budding. If you wish the plants to develop more callus, just before setting them out you can use a bottom heat of 65° to 70°F. for about a month. A few plants such as apples, pears, and plums are quite difficult, and with these you might want to continue the bottom heat for up to two months. Keep the tops up in the cool air, however, and see that they do not get soggy. Cuttings from fruit plants like these are often cut much longer than others, and for bottom heating are inserted into deep sawdust or sand. But such methods probably work best in rainy climates like the Pacific Northwest or England, where cuttings so treated sometimes root in a mere six weeks. Recent experiments with starting treated cuttings in polyethylene bags have been very successful, and if you want to store them for future use, you need heat of only 50°F.

Putting Cuttings Outdoors If the climate is suitable, there is no reason why you can't put hardwood cuttings in a sheltered place outdoors from the beginning. Choose a spot in a warm corner of the garden, where the drainage is very good. (Otherwise make a raised bed for the storage of the cuttings, for you do not want them to get wet and rot.) Open up a shallow trench with straight vertical sides, and stand the hardwood cuttings on end in it. Then push the soil firmly against the stems, leaving the top quarter of each uncovered. This can be done as soon as you take the cuttings, so have the trench ready before you start cutting.

The soil in the trench should be sandy, with both peat and sand worked in. See that it is well worked over. If the medium you use is pure sharp sand, transplant the cuttings when rooted so they can get sustenance. Herbaceous cuttings can be put into pots and moved to a shaded place where there is good ventilation. If they were taken in the spring, they will be ready to set out in the garden by autumn. Shrub cuttings that you put out must be kept moist until well established. A good mulch is needed. Some plants will do better if potted and sunk in soil in the garden for further development before being planted in their permanent place in the garden.

Repeated Transplanting It is undoubtedly a good idea to move plants made from hardwood cuttings several times before setting them out so they will develop a good mass of fibrous roots. In many nurseries that is the custom, and it certainly is a good idea to bring the plants on to a sturdiness that is desirable.

Forcing Another nursery practice for plants like fuchsias, hydrangeas, and roses is to pot up the stock plants and move them to the greenhouse, where they can be forced to put forth new, vigorous, rapidly growing shoots. Head them back to encourage lateral shoots to grow, and then these lateral shoots can be used for the cuttings. Outdoors, with your own plants and shrubs, you can always cut back old woody growth to force new growth suitable for cutting material. But remember to keep down the nitrogen supply so that the new

The pot-within-a-pot arrangement works well in keeping cuttings moist as their new roots develop. Place a small clay pot that has had its drainage hole plugged up in a larger pot, and fill the area between the pots with rooting medium (1). Fill the smaller pot with water (2) and insert the cuttings in the rooting medium (3). Water from the smaller pot will seep through its clay sides into the rooting medium, preventing it from drying out.

growth will not be soft and juicy and unsuitable for making cuttings.

Using a Double Pot   Another excellent method for cuttings is to put them around the rim of a clay pot and to insert a smaller, corked clay pot in the center of the soil. Keep this inner small pot filled with water, and let it merely seep through the clay into the sand or sand and peat in the rooting mixture. When capillary action is established, there will be a very slow, continuous movement of water into the sand. Be sure to remove any glass or plastic covering at frequent intervals to check fungus growth or black rot.

Whether you use the inner pot or not, it is always a good idea to insert cuttings near the rim of the clay pot. This may be partly because there is more air for the roots out near the porous clay, and partly because there is a continuous healthy passage of moisture out through the clay when the air is drier than the contents of the pot. In other words, normal soil activity of water and air is greater out near the rim than in the less airy, perhaps more waterlogged interior of the pot. And remember that small cuttings, like small seedlings, like to grow near each other.

Making Effective Rooting and Potting Mixtures   As I explained in chapter 1, gardeners have used many different mixtures for successful cuttings. These range from simple moist sphagnum moss to combinations of coarse sand and loam, peat moss, vermiculite, perlite, or several of these together. Fine sand and salty beach sand are no good at all. Whatever mixture you use, your primary aims are scratchiness, good aeration, and good moisture retention. When there is loam—even pasturized loam—in the rooting mixture, you run more danger of infection from fungus disease; on the other hand, unlike the nutritionless other materials, loam does offer nutrients. But since roots will form in these sterile media, all you need to do is keep an eye on the cuttings and pot them up in a fertile medium as soon as you see that they have roots 1 to 1½ inches long.

Since you don't want to subject the plant to too much of a shock in transition, it is wise to move from a mixture of equal parts of two materials to a mixture containing three, or from three to four parts, by adding loam. For example, you can add 1 part of rich loam to 1 part each of peat moss and sand; or add 1 part of rich loam to equal parts of vermiculite, peat moss, and sand. If you are taking cuttings from acid-loving plants, use equal parts of sand, acid peat, leaf mold, and Michigan peat.

As an aid in putting together your rooting and potting mixtures for stem cuttings, keep in mind that:

SAND—is clean, easily sterilized, and capable of holding some water, especially if it's somewhat fine. Coarse builder's sand is best to use, however, because the main function of this medium is to assist drainage. Though sand is a good rooting medium, roots growing in pure sand sometimes get long and unbranched. If this happens, begin to add soil.

PEAT MOSS—should be used in combination with some other medium. Its main virtue is its water-holding capacity. Peat has a few nutrients, but not many. When misting cuttings, beware of using more than 1 part of peat moss to 2 or 3 parts of a less porous medium, or your mixture will get too soggy. Imported peat is acid; Michigan peat is not.

SPHAGNUM MOSS—is natural acid bog moss. It is almost sterile and reportedly somewhat antibiotic. This medium is very good at holding water, as it does in nature. It is usually shredded, but not necessarily. Its pH of 3.5 inhibits the damping-off fungus.

VERMICULITE—comes from a mica-like mineral that has been expanded by heat. It has a pH of about 7. Very water-absorbent, it can hold nutrients in reserve by slow release. It will provide plants with the magnesium and potassium they need while rooting. Vermiculite is completely sterile because it has been heated to 2,000°F. during the expansion process. Use #2 for cuttings; but for seeds use #4.

PERLITE—will hold water up to 3 times its weight. It also is sterile, but has no nutrients. Of volcanic origin, it is heated to 1,400°F. for expansion.

COMPOST—offers many recycled trace elements and an advantageous carbon/nitrogen ratio. Because it contains mellow humus, it holds water well. Compost is usually mixed with soil, but since the compost is already pasteurized during the hot period of the composting process, the pathogens are probably gone.

SAWDUST, SHREDDED BARK, etc.—are usually from conifers such as pines, firs, cedars, and redwoods. These materials all hold water, decay slowly, and need supplementing with a high-nitrogen material such as blood meal, soy meal, cottonseed meal, alfalfa meal or the like if the cuttings are to stay in the material for very long.

SOIL—is composed of many things in gaseous and liquid, as well as solid, states. It contains both inorganic and organic substances. The latter can be alive or dead and in some stage of decaying into humus. Textures of the component mineral parts vary from pebbles and gravel to sand and silt and very fine particles of clay. The organic parts may be sticky, prickly, hard, or soft. Soil is a rich, normal, and indispensable medium for plant growth. But when used for cuttings which have not yet rooted or for roots not yet sprouted, it may cause difficulties unless it is pasteurized. Unusually hard-to-root materials are benefitted by being rooted in soil; this is especially true of stem (and root) cuttings of deciduous hardwoods. Use sandy loam for these hardwood cuttings. With softwood cuttings, use twice as much sand as soil in the mixture.

Other Kinds of Cuttings

Root Cuttings

Cuttings made from roots involve a somewhat more delicate process, partly because roots should not be exposed to the air for any length of time. When well-done and well-planted, root cuttings usually send out *adventitious buds*. (This term is used to describe buds that arise from somewhere

other than leaf axils.) Sometimes this happens when you merely wound a root. Also sometimes a callus is formed on the root and the bud comes from that. And sometimes, especially if the roots are young, the bud comes from an area near the cambium. After sending out this adventitious shoot, the new plant then sends out roots, if there are enough growth substances present to bring this about.

If you work fast, you can start making new cuttings from the new shoots, and you can also start again with root cuttings when the new shoots send out their first roots. When the new roots come, they may be adventitious—growing from stems or leaves—or may come instead from old roots with root initials on them. (Here auxins play a part.) In some instances, however, the root may send up one good adventitious shoot but develop no more roots. This kind of cutting will die.

It is best to use roots from young stock, and to cut the pieces in late winter or early spring when the roots are still well supplied with stored food. Do not take root cuttings after this food has been spent in sending up new spring growth. One good reason for using young stock—aside from the basic biological advantages—is that it is easier to get at than the roots on bigger, older plants. Just trim the roots at transplanting time, for instance, and start new plants. Be sure when you do it to replant the pieces of root right side up; that is, plant uppermost the end of the root that was nearest the plant when you took the piece off. It helps to use one sort of cut at the

To propagate a raspberry bush by root cuttings, dig up a young plant (1) and snip off 1- or 2-inch pieces of the main root (2) and (3). Prepare a flat of fine sand, lay each cutting on the sand (4) and cover them with more sand (5). To give the cuttings something to sustain them as they grow, you can then sprinkle some pasteurized loam on top of the sand (6).

upper end and another at the lower in order to distinguish which end is up. (Auxins or root regulators are not used with root cuttings, for they tend to inhibit shoot and bud formation, and these are what you want.)

Some plants have fine, small roots that are especially hard to handle and care for. Work rapidly and gently. Plant 1- to 2-inch lengths at once in very fine sand, or better, lay them on sand and then carefully cover them with more fine sand. If you wish, you can add some fine pasteurized loam at this time, so that when the cuttings are established, you do not have to move them again from a sterile to a nutritious medium. Water gently and cover with glass or polyethylene.

Fleshy Roots   Roots that are fleshy rather than fine and small can be handled more easily and cut in 2- to 3-inch lengths. After you plant pieces vertically in sandy soil, they will form adventitious roots quite rapidly. Then you can move them to the garden, or if a house plant, to a new pot. If you're in doubt about the sterility of the root, start it in peat and sand and move it later. House plants such as geraniums (*Pelargonium* spp.), Kafir-lily (*Clivia miniata*), glorybower (*Clerodendrum thomsoniae* or *C. speciosissimum*) and others sprout easily when cut and put in peat and sand. You can do this with bananas, too.

If you propagate fleshy roots out-of-doors, take cuttings 3 to 6 inches long, tie in bundles and store in damp sand or peat moss for 3 weeks at 40°F. Then untie and plant 2 or 3 inches apart with the tops of the cuttings just barely below the surface of the soil, or even just level with it.

Tubers of caladiums, begonias, and potatoes do better if left exposed to the air for 2 or 3 days to form a callus by the action of traumatin, a wound-stuff hormone. (See also Tubers in chapter 4 on Division.)

Other Root Cuttings   Suitable for root cuttings are such perennials as oriental poppy (*Papaver orientale*), bleeding heart (*Dicentra spectabilis*), California poppy (*Eschscholtzia californica*), leadwort (*Ceratostigma plumbaginoides*), and phlox (*Phlox paniculata*). Several roses such as the very hardy

Fleshy root cuttings of the oriental poppy are planted upright in a sandy rooting medium for propagating.

meadow rose (*Rosa blanda*), the shiny-leaved, red-flowered shining rose (*R. nitida*), and the Virginia rose are propagated by root cuttings; so are other shrubs such as the sweet fern (*Comptonia peregrina*), bayberry (*Myrica pensylvanica*), lilac (*Syringa vulgaris*), various raspberries and blackberries, and the rose-acacia (*Robinia hispida*).

Trees, too, can be suitable for root cuttings. Examples are the bottle-brush buckeye (*Aesculus parviflora*), tree-of-heaven (*Ailanthus altissima*), Japanese pagoda tree (*Sophora japonica*), sumacs and aspens, and the lovely goldenrain tree (*Koelreuteria paniculata*). (See Part II for further suggestions.)

A few kinds of plants develop adventitious buds on their leaves, and these may be used for leaf cuttings or leaf propagation. This happens on the house plant kalanchoe. All you have to do is peg down a leaf along its edge on a rooting medium and wait to see the new plants appear along the edge. Rex begonia leaves can also be propagated this way, but peg down the veins after slitting them on the underside of the leaf to make the wound. Keep the leaf in contact with the rooting medium, keep it moist, prevent fungal attack, and watch for the new plants and for the old leaf to disintegrate.

The house plant generally known as snake plant (*Sansevieria* spp.), can be rooted from the base of a leaf or from pieces of leaf which are inserted into the medium for rooting. It is advisable to cut the top and bottom in different ways so you are sure to insert the bottom cut in the rooting medium. This piece will send out a new plant but will not become part of that new plant. If you use variegated leaves they will not come true to type. Division is the only way to reproduce the variegation, for it derives from the root cells as well as the areas from which both the solid green and the paler, whitish color come.

Sometimes propagators cut triangular pieces of fibrous-rooted begonias' mature leaves so that each piece contains a big vein and the outer edge is taken off. Then these cuttings are inserted vertically in sand, with the pointed end down. The new plant comes from the bottom of the large vein.

Leaf Cuttings

To propagate a begonia from a leaf cutting, make several slits across the veins on the underside of the leaf (1). Lay the leaf on the medium, vein side down, and then peg down the veins through these slits (2).

Snake plant (*Sansevieria* spp.) roots easily from leaf cuttings. Cut a leaf into about 3-inch pieces (1) and plant them, bottom cut down, in a rooting medium (2). Each leaf cutting will send out a new plant, and once this new plant is established the original leaf cutting can be discarded and the new plant repotted.

People also cut small pieces from large-leaved begonias and simply put them on moistened paper under glass and wait for new growth.

Another very simple method can be practiced with peperomias, most gesneriads, and especially with African violets *(Saintpaulia* spp.). Just take the leaf and its petiole and insert it into the rooting medium (the leaf blade only or just a part of the leaf can also be used). When the petiole is left on, the new plant will come from the base.

It is best to set the cutting into the medium so that the leaf blade is almost touching it. Keep it warm and moist. But if there is any chance of fungus, use some charcoal in the mixture. If you see any signs of fungal growth, clean it off and air it at once. If the cutting needs propping, use sterilized pebbles.

You can practice no-work propagation with plants like the pickaback, or piggyback (*Tolmiea menziesii*). New plants come wherever a leaf blade touches the soil or rooting medium. All you have to do is wait; when the new little plant is 2 inches tall, cut it off and there you are! (See Part II for other plants that root from leaves, and for scales on bulb plants, see the section on Bulbs in chapter 4.)

Leaf-Bud Cuttings   You can be even more successful sometimes with cuttings made with both a leaf and the bud which grows in the axil, or angle, of the leaf. Make an oval cut into the stem, taking a flap of bark above the bud and leaf stalk, as well as one below. If the stem is small, you can just cut through it a little above and below the bud and leaf stalk. Usually you cut off the leaf blade, leaving the petiole (which makes a handy handle to work with). Then push the cutting down into a sandy rooting medium, far enough to cover the bud ½ to 1 inch; press it firmly enough to keep the cutting from toppling.

Because each node can be used for a cutting, with this method you can get about twice as many new plants from the same amount of material as you'd get by taking straight cuttings. The new shoots come readily from the already formed buds. Bottom heat and mist-sprays help. High humidity is essential.

A leaf bud cutting has been made here by cutting off the bud and leaf stalk. The leaf blade was cut off, leaving just the petiole (1). Then the stem of the cutting is inserted in a rooting medium so that the bud is completely buried (2).

Leaf-bud cuttings are used for such species as the bramble fruits, with the exception of red raspberries, and for camellias and many of the tropical shrubs and plants grown in greenhouses. Lemons can also be propagated in this way.

Shoots springing up from roots, stems, or underground stems are called *suckers*. The term is also applied to any shoots coming from below the point of union of a graft, and the highly desirable practice of removing such shoots is called "suckering." On the other hand, water sprouts, which are also often referred to as suckers, come from a latent bud of a

Suckers

stem already several years old, sometimes even from the main branches or trunk of a tree. We had many appear a few years ago on one of our pear trees after the tree had been damaged during a bad windstorm just when the limbs were laden with fruit and too heavy to withstand the wind.

Easily cut off, suckers are usually quite easy to root, for they arise from adventitious buds. They are also useful when the propagator wants to get new plants with juvenile characteristics, including good rooting. The suckers sent up from underground roots, stumps, or stem parts are sometimes used as cuttings to grow stock plants for grafting. They are also used to carry on the stock of a good new seedling, and they are the main source of materials for propagating bananas. And, of course, suckering is the natural mode of propagation of raspberries. But seedlings are generally considered better for propagation because a sucker has the tendency to pass on its suckering habit and be a nuisance. What's more, as with water sprouts, too much vigor goes into the suckers and not into the fruiting branches of the plant.

Red raspberries send up suckers from underground parts. If you want to encourage new suckers to make new plants, you can mound up the base of the plant a little with good loam. Sometimes people merely pull the new plants or suckers away from the mother plant, but it is a better practice to dig up the whole plant and cut away the suckering parts for replanting. Replant the old part, too, and mound it again for next year's propagation. Blackberries can also be propagated by cuttings of suckers, though root cuttings are more usual with this species. (Black raspberries are propagated by tip layering. See that section in chapter 3.)

Lilacs are notorious for sending up suckers. These shoots can be easily cut off and used for making new plants, but be sure that the root is of the same stock as the upper plant, otherwise you might end up with a privet plant because the sucker grew from the stock to which the lilac was grafted.

Japanese quince and snowberries are also easy to renew from suckers.

This photograph shows suckers growing around a crab apple tree. Such shoots should be cut off to improve the tree and used like hardwood cuttings for propagating.

Some Comments

I never cease to wonder at the marvelous fact that a stem or any other part of a plant has stores of enough food and energy to keep a cutting going until it is rejuvenated and ready to start new growth. I also find it wonderful that in asexual propagation the new plant is so similar to its mother plant. Cell division in the new clones (or new asexually propagated plants) is always of the sort that passes on the same number of chromosomes and never of the sort that divides the number of chromosomes in two, as in sexual reproduction. Plant breeders and all gardeners have found the transfer of identical characteristics from mother to daughter plant in asexual propagation of great benefit, of course. Hybridizers and orchardists rely on this transfer. Seedless grapes, figs, bananas, and many citrus species could never have developed without the resource of asexual propagation. It also adds to the pleasure of making cuttings, I always feel, to think of how economical a way this is to get new plants, and how useful a way, too, when you can save twigs to root during the process of clipping your hedge or pruning your roses. Yet, whatever your pleasure, the deepest excitement about working with cuttings is that you make new plants from old and take part in the wonderful and mysterious event of rejuvenation.

Chapter 3

Layering

Layering is one of the easiest methods of vegetative plant reproduction involving the separation of a part of a mother plant to form a new plant. It is very suitable for amateurs, for you merely push a more or less flexible branch down into the soil, wound it, cover it, peg it down, and wait for the roots to form. Or you wound and wrap up a stem and wait for roots to appear under the wrapping in the process called *air layering*. In both instances, the moisture, lack of light, and supply of air all contribute to the formation of adventitious roots. Whether the wound is a slit or a girdling of the stem done by removing a ring of bark, it blocks the passage of carbohydrates and hormones and helps to initiate callus formation.

Some plants bend the tips of some of their branches down into the soil naturally. Fruits that layer themselves include black and purple raspberries as well as gooseberries, currants, and trailing blackberries. Towards the end of the season the tip of the current year's growth bends toward the ground on its own or with the gardener's help. The tip grows downward, then curves up again to produce a sharp bend in the stem. The lower part of the bend touches the moist earth (or is slightly covered with soil) and roots begin to form. In plants which do

Tip Layering

this, the canes are vegetative the first year and bear fruit the second year, so they are actually biennial. After two years they should be pruned back.

You can tell when the canes of these tip layerers are getting ready to bend down because their terminal leaves get small and the tips look long and ratty. Help them along and bend them down soon; if you do, that will encourage new terminal buds to form and will keep the tips from growing and growing. Use a trowel to make the hole for each cane, slicing so the far side of the hole is straight and the near side slopes towards the parent plant. Lay the cane against the sloping side and press it in when you cover it with soil. This will hold back the tip growth and encourage lots of roots. A good humusy soil—or one that is not hard and clayey—also encourages good root growth. Then will come the nice, vigorous young shoot, and you have a new plant. Stake it if it wobbles.

Plants used for tip layering are best segregated in one plot and spaced up to 12 feet from one another to make room for future layering. Cut all the canes down to within 9 inches of the ground in the spring, and let new shoots come out from the base. When the cane gets to be 20 or 30 inches long, pinch off 3 to 4 inches of the tip to encourage lateral shoots useful for tip layering (or for the next year's fruit, if you let them go full cycle). Dig the new tip plants at the end of the season and keep 6 inches of the old cane for a handle to the young shoot. Be careful of the mass of roots and the terminal bud, all of which are rather delicate to handle safely. Work fast and replant fast, too, for the new little plants will dry out if you don't. Some people do this in late fall, others do it in early spring. You'll find that the new canes grow very fast the first season.

Simple layering can be used to propagate low-growing plants with soft or semi-woody branches. An evergreen azalea is being propagated here with a variation on the simple layering method described in the text. (1) Select a low branch that can be easily bent to the ground. Nick that part of the stem that will be in contact with the ground. (2) Loosen the soil where the branch is to be placed and *(cont.)*

Simple Layering

The process of simple layering is somewhat similar to tip layering, but the ends of the branches are left aboveground for 6 inches or more. Also, their leaves are left on, except for the area 6 to 12 inches below the tip where you remove the leaves before burying that part of the branch in moist, loose soil. Simple layering is also somewhat similar to natural propagation by runners, where the plant strikes roots where the runner hits the soil.

With the layering you do yourself, wound the stem by making a slanting slit in it where you are going to bury it. Sometimes a pebble or toothpick or little twig is inserted into the slit to help it stay open, but the degree of the bend should keep it open in most cases. It will always stay open if you give it a good twist and insert it so the tongue of the slip points down and the end of the stem sticks upright. Bury this wounded part of the stem under 3 to 6 inches of soil. Use a stone to weight it down if the stem won't stay down by itself. To keep it moist is sometimes quite a job, but if layers are in the shade, and if you mix some peat moss into the soil and pay attention to watering, the layered stems will take root and survive. Remember, however, that even one day of drying out will probably kill the tender young roots.

Professional propagators use simple layering for multiplying rhododendrons, which are notoriously difficult to propagate from cuttings. They bend down all the branches, right around the bush, though the bushes they use have already been cut back and headed down very low to get all these young, strong shoots near the ground. They put down the layers in March or early April, as soon as the soil is warm, using the previous year's growth. All flower buds are removed, as they should be when layering, to encourage root and vegetative growth. Commercial growers leave the new plants in the ground for a year, severing them when the plants go dormant, but letting them stay in the ground as independent plants until they are strong enough to transplant. This is a good practice for anyone layering shrubs such as forsythia or little willow or hawthorn trees. It can be quite a shock for a new small plant to be severed from the mother plant and given the upset of transplanting all at the same time. Plants more suitable for layering than for taking cuttings include magnolias, lilacs, leucothoe, photinia, silk-tassel, most procumbent shrubs, and very small shrubs such as rose daphne (*Daphne cneorum*).

In serpentine layering, instead of dipping the branch down once into the soil, you dip it down several times. It is suitable,

be sure that it is rich and humusy. (3) Bend the branch to the ground and cover with about 1 inch of soil. Place a rock on top of the buried part of the branch. The rock must be heavy enough to hold the branch firmly in the soil. Layered branches should be left in place at least a year. After that time, remove the rock and check root growth. If there are many roots the branch won't budge. When the plant is dormant (fall is a good time) cut the well-rooted layer from the mother plant just below the new root system and when the rooted cutting has established itself as an independent plant in the spring, transplant it. Transplanting is best done in spring so that the layer will have a full season to adjust to its new location.

Serpentine Layering

therefore, for honeysuckles of various varieties, jasmines, and for vines. The viny nasturtium (*Tropaeolum majus*) also can be propagated by serpentine layering; others possible to use include clock vine, potato vine, scarlet runner beans, sweet peas, morning glories, ivies, cup-and-saucer vine, and clematis.

Continuous Layering

To do continuous layering, simply lay down a whole branch or whip and peg it to the ground at intervals, in a continuous line. As young shoots grow from the points of contact, you should mound them up and up until the soil covers 6 or 8 inches of each. The pegging down of the stem can be done in the fall, and the mounding the following season. At the end of that second season the rooted shoots may be removed and transplanted. Leave one young shoot near the base of each layered branch to grow up the next year.

This approach is sometimes called the *etiolation method* because the shoots are never permitted exposure to light. Roots grow well from etiolated plant materials.

Trench Layering

A variation on continuous layering, trench layering, can be used for a whole little tree or a branched limb. The custom is to start tilting the little tree the year before, by setting it in the ground at an angle of 30° to 45°. (If done in rows, keep the rows 4 feet apart.) Then cut back the tree and allow it to grow for a year.

After the little tree has grown through the first winter following tilting, push it down and bury it in a trench about 2 inches deep. Be careful to dig this wide enough to take care of the side branches. Soil is a good medium to use, but you can substitute or mix in peat moss, wood shavings, or sawdust if you wish.

Soon the new shoots will start to come up from the buried stems. Now add more soil or mixture, so that you insure that at least 2 or 3 inches at the base of the new shoots will be in the dark. Let the shoots grow perhaps 3 or 4 inches more, and again cover up to half of the newly exposed part of the shoot.

Keep mounding until mid-summer, so that at last about 7 or 8 inches are covered.

After the plants have become dormant in the fall, you can begin to separate the newly rooted shoots and prepare to replant them. Do not cut off shoots which have failed to root. If you're in doubt, wait until the next spring to carry out this whole last operation. If you plan to go in for continuous trenching by pegging down some of the newly formed shoots as new layers, it is probably wisest to wait for spring, anyway.

This trenching method is most often used on difficult fruit trees. It can also be used for vines or other long flexible stems, but it is wisest to leave the tip out. When used on dwarf shrubs, such as Labrador tea (*Ledum groenlandicum*), salal (*Gaultheria shallon*), wintergreen, or blueberries, the method is also called *drop layering*. The little plants are dropped into the soil on the slant so that the young tips near the top of the branches are half-covered with soil when the layering has been done.

Mound or Stool Layering

The variation known as mound or stool layering is especially suitable for plants which will send up a whole group of new shoots when the plants are cut back nearly to the ground. (The stump from which these shoots grow is called a *stool*.) If the parent stock is well situated on a bed of loose sandy loam established the year before mounding, this method of propagation can be continued year after year on the same plants.

To get started, put in a small, well-rooted new shoot from a previous layering of quince, gooseberry, currant, or certain apple stocks. Start it in a shallow trench, piling up soil and filling the trench as the shoot grows the first season. After the period of winter dormancy, cut it back to the ground and wait until the new shoots are 4 or 5 inches high. Then begin mounding, using soil or sawdust, covering half the height each time until you have made a mound of 6 to 8 inches. At the end of the summer all the new shoots will be rooted, unless you have had a disaster. (One disaster is for the little roots to get too much heat in the mound.) Remove the new shoots to transplant, and leave the mother stool to send up more shoots the following spring. The purpose of the trench the first year is to put the plant low enough so that the new shoots will also start low. If the shoots seem very crowded, spread them out in the mounding medium as well as you can.

Air Layering

(1) To air layer a leggy house plant such as this *Dracaena marginata* to get a shorter, more compact plant, start by making a slanting incision with a sharp, clean knife, preferably just below a leaf scar, as is being done here. (2) Then insert a small piece of wood into the wound to keep it open
(cont.)

Now that we have polyethylene film to keep the wounded section of the stem damp, I sometimes think air layering is the very easiest of all methods of propagation. Because the Chinese practiced this method centuries ago, it is sometimes called *Chinese layering*. Before we had polyethylene film as a protection from drying out, air layering was mostly practiced in greenhouses. There the wounded stem, wrapped in sphagnum moss, was kept moist by means of high humidity and constant care. Air layering was also practiced outdoors in the tropics. But the heat in both situations was a challenge, with the moss and its paper wrapping threatening to dry out several times a day.

Now air layering is done by many amateurs, both to propagate leggy house plants and to improve their looks; it's also done to increase the propagator's stock of outdoor plants.

For this method, work with young growth on partly hardened wood in the late summer; or use the previous year's growth in spring. Wait until there are several active leaves on the stem so there will be a good food supply. On house plants do your air layering after the plant has come out of its rest period and several new leaves have appeared. Cut a longitu-

dinal wound of about 2 inches which enters the stem nearly to the core. This cut should be made about 6 to 14 inches from the tip. Next insert a pebble, little stick, or a small wad of damp sphagnum moss to keep it open.

You can also try girdling the stem by making two cuts around the stem ½- to 1-inch apart and scraping away the bark and cambium in the area between the cuts. This will help to encourage callus and root development and retard healing. You can add indoleacetic acid and talc if you wish.

However you make the wound, you must wrap the wounded area with a big handful of moist sphagnum moss. (If your hands are small, use 2 handfuls.) The best way to get the right degree of moistness is to soak the moss first for 2 or 3 hours, then wring it out until it feels like a moist sponge. The moss must not be really wet for that might cause rot. Next, take a square of polyethylene—say 8-by-10 inches—and wrap it around the moss so that it is completely enclosed and so that when you tie it, little air can get in and little or no water vapor get out. Where you have to overlap the film, tuck it in at the bottom of the wrapping to help keep out rain. Over the film you can apply an adhesive tape, such as the waterproof kind that electricians use (Scotch Electrical Tape® #33). Start the taping on the bare stem, and then wrap it around and around until it seals the top of the film. And do the same at the bottom. Be particularly careful at the top, again in case of rain. You do not want a soggy mess of moss around

and help roots form. (3) Wrap the wound with moist, unmilled sphagnum moss or sheet moss, mixed with vermiculite. Peat moss soaked in warm water can also be used. (4) Tie the wrapping with plastic ribbon or twine to keep it tightly in contact with the stem area. (5) Then cut the bottom out of a plastic bag and slip it over the top of the plant and wrap it over the wound area. This will preserve the mosture and protect the wound. Add water from the top if the moss dries out, but never let the moss get soggy. Give it good light, but not direct sun. (6) Stake the plant if it's as leggy as this one to give it extra support. (7) After 4 weeks remove the plastic and check to see if roots have formed. Do not consider cutting the layer from the mother plant until the moss is thoroughly filled with roots. And do not remove moss from delicate roots, for that would tear them. When the layer is ready for cutting (this *Dracaena* took 8 weeks to develop root growth heavy enough to justify potting the layer), untie the ribbon and hold the moss in place. (8) Then cut the stem just below the area where the new roots have formed. (9) Plant the new *Dracaena* by burying the root ball an inch below the soil surface Discard the old stem. Frequent watering and a cool location are advisable for newly layered plants, though *Dracaenas* should dry out between waterings. Soil recommended for *Dracaenas* is 3 parts coarse compost to 1 part sand.

the stem which will completely cut off air from developing roots and stop the whole process.

As usual, the stem to use should have about the same diameter as a pencil. Like the ideal season for making an air layer, this ideal size is subject to a certain amount of variation. That is, you can use older, larger wood and get good results. Experiment and see what you can discover. You'll probably have good luck with pencil-sized stems of such widely varying plants as the dove tree (*Davidia involucrata vilmoriniana*), the charming Franklinia (*Gordonia alatamaha*), and crab apples, maples, poplars, viburnums, and catalpas as well as wisteria vines. (And bad luck with oaks, hawthorns, lilacs, other viburnums, and other apples.)

The big hazard—as with cuttings started in water—is that the rooted air layer will not adapt well once it is transferred to soil. It would be worth your while to soak the roots of the newly separated plant in warm or lukewarm water, then ease it into a halfway medium like vermiculite and peat moss dampened in compost tea. Finally, add soil gradually in order to allow the roots to adapt.

Pot laying is almost always successful because the stem is still attached to the mother plant while it is forming its own roots.

Pot Layering

Each of the half-dozen runners and new little plants of this spider plant can be layered into a pot of rooting mixture and left to grow into a new plant. Here one runner has already been put into the pot.

Layering in a pot is often referred to as a variation of air layering, but it really is a combination of air layering and simple layering. You treat the stem and wound it just as you do in air layering, but instead of wrapping up the wounded area, you bend it down and bury it in a rooting mixture in a clay pot. This should be large enough to hold the entire area, and wide enough to accommodate the emerging tip of the branch, which you leave out in the air. The advantage over taking cuttings is that the newly forming plant is still attached to the mother plant, getting sustenance. There's also a benefit lacking in air layering: the new roots get used to soil or the potting medium as they are forming. (As usual, keep things sterile.)

Making Use of Natural Propagating Methods in Layering

You can use a variation on the natural propagating method involving runners to help establish very stong new daughter plants. Simply ease the new plants very gently from the earth where they are taking root and replant them in a flowerpot well supplied with rich loam. In many instances it is best to sink the pot in the ground so as not to subject the prostrate stem to too much stress. You can peg it down, too, but do not sever it yet.

This method is successful with strawberries, with the house plant spider plant (*Chlorophytum elatum*) and with apostle plants (*Neomarica northiana*). The various hens-and-chickens (*Sempervivum* spp.) can be propagated this way, too, if you ever find any with long enough stems. (If not, you'll just have to treat them as offsets, gently separating the daughter plants from the mother plant before replanting or repotting them.) Bromeliads of several kinds are more amenable to transplanting before separation, for the connecting stem is usually a good deal longer. This is also true of pickaback plants.

The mints, many weeds, and the red osier dogwood (*Cornus stolonifera*) send out special modified stems called *stolons*. Unlike runners, which usually start near the base of the mother plant, stolons begin above the soil surface, then bend over to put down roots. They grow horizontally along the ground and send up new stalks at intervals. Though I do like potatoes, and I'm glad they have stolons which produce the tubers I like to bake, I usually think of plants with stolons as pests, even the mint, though I am glad enough to have it growing here and there in our backyard. You don't have to layer mints because they layer themselves so efficiently. You can, however, treat them as layers and cut new plants away from the mother plant to put in other parts of the garden. You could have mint at every corner, I should think, once you get a few plants started. Stoloniferous plants must be a boon to nurseries—all those new plants with only that little work.

The offsets on aloes and lilies are charming. These are like new shoots or short lateral branches coming from the base of a main stem. Unlike runners and most stolons, an offset produces only one new plant at its end and not a series, so offsets do not spread all over as stolons and runners do. I was delighted one day last year to see the tiny offsets coming from the soil beside my medicine plant (*Aloe vera*). They kept coming and multiplying and in a few months there were plants-a-

Runners

Hens-and-chickens propagate themselves readily by runners, forming a lovely ground cover, as seen in this close-up here.

Stolons

Here mint shoots are growing from an underground rooted stolon. To propagate, merely cut the stolon between each shoot and replant.

Offsets

plenty. When offsets like this are well rooted, you can cut them away from the mother plant and pot them up by themselves. If you happen to cut them off before they are well rooted, it is too late, obviously, to put them back. The thing to do is put them in a rooting medium with a little bottom heat and treat them like leafy softwood cuttings. Other pleasant plants with offsets are tender lilies like lily-of-the-Nile (*Agapanthus africanus*), rain-lily (*Cooperia* or *Zephyranthes atamasco*), blood-lily (*Haemanthus multiflorus*), and Guernsey-lily (*Nerine sarniensis*). Sometimes the growths on lilies are merely new bulblets, though botanically they are called offsets, too. (See Bulbs in chapter 4.)

The medicine plant (*Aloe vera*) that you keep in the kitchen to break off and put on burns can be multiplied by lifting the plant and removing offsets. Here three offsets are being separated from the mother plant. Be sure the offsets you choose have roots.

Maybe one reason for this nomenclature is the biological activity which leads to all these runners and other extensions from a plant. Tipped with an apical meristem, the new shoots come from the cambium, the thin-walled parenchyma cells within the phloem, or an area just outside the phloem. Once formed, they grow out through the cortex and outermost layer of tissue in search of the moisture and oxygen they need. Offsets often develop fast. Propagated lily offsets, for example, come to bloom as good robust plants long before you could expect any such plants from seed; with the vegetative process the plants hurry on towards flowering in a rhythm quite different from the slower evolving rhythm of seed to plant to flower to seed occurring in sexual reproduction in plants.

Special Examples of Layering

One good example of layering I learned about from an article in *Organic Gardening and Farming* is that of tomatoes. The plan is to encourage the suckers which grow up from the axils of the side shoots and not to remove them as many tomato growers do. When they get to be 12 to 15 inches long, bend them over and cover most of each sucker with a damp, rich mulch, leaving a tip of 3 inches uncovered. As with young tomato plants just set out, it is best to surround the growing layer with a collar to keep cutworms from invading and eating off the tip. The roots form well in about 2 weeks, when you can sever the new plant and set it in the row with the rest of your plants. (If you make many layers, you will probably have enough for a whole new row.) When these new plants get suckers, you can layer them too. And at the end of the season, do it once more and get nice new plants to pot up for growing indoors until Christmas.

Tomatoes

One of the good methods for propagating wild flowers is to layer those which respond well to this treatment. Both simple layering and tip layering can be practiced, as well as mounding and serpentine layering if the plant is a vine or has long supple stems. Like cultivated plants, creeping plants you layer will root in just a few weeks. Then you can move them, either to a new situation in their native habitat, or into a nursery for further growth and further propagation. With wild plants it is usually a good idea to put down a thin sprinkling of sand before you introduce the moist peat and sand mixture or vermiculite and sand which you use for the rooting mixture. It is wise to do this because the leaf mold may not be as sterile as it should for the health of the adventitious roots once they begin to emerge and grow. Sand beneath also helps drainage and prevents any tendency to rot, or at least inhibits it.

Wild Flowers

Though it is not a very frequently used method, layering of garden perennials or biennials is sometimes quite an easy way to propagate plants which do not respond well to propagation by cuttings. Carnations, for example, are quite easy to handle. Surround the plant with sand and peat some day in July or August. Layer whatever 3 or 4 stems you wish, and peg the stems down with a hairpin or bent wire. As with

Garden Flowers

other layering, results will be vastly better if you slit the stem and make a wound first. You can cut to almost halfway through the stem and dab on a little indoleacetic acid in talc. Mist-spray to keep the rooting medium moist. You also can build a little plastic tent (with air holes punched in it) to help hold moisture. But be sure you inspect it regularly—especially on hot days—to see that the air is not too hot and causing wilt, or too muggy and causing fungus trouble. When rooting takes place—probably by autumn—you can pot up the new plants and have them for winter decoration in the house, or let them winter over in a cold frame or greenhouse until planting out time the following spring.

Other perennials can be layered by mounding. With bellflowers (*Campanula* spp.), for example, just work good sandy soil around the base of the stems and wait for roots to come out at the base. Then in fall you can get a good many new plants by division. (See Simple Division and Crown Division in chapter 4.)

Roses

Except for hybrid tea roses, which are grafted, roses respond well to several layering methods. You can do simple layering by slitting the stem partway, bending down the cane, and covering the slit part with moist peat and soil. If this is done some time in July or August, the layers will take root by autumn and can be cut away from the mother plant and put where you want a new rose in the garden. Climbers respond especially well to simple layering.

Rambler roses can be propagated by tip layering. Follow the steps described above for bramble fruits, and remember to bury the rose tips just about 4 inches. Wait until the new plant is well rooted before you transplant it.

Suckers on roses can simply be cut away from the old plant when it is dormant and moved to a new situation, where they can be treated as cuttings. They can also be moved to a cold frame and wintered over in 6 inches of peat and sand. Cane roses can be air layered.

Bulbous Plants

Stem layers of certain bulbous plants can be made in the hope of encouraging a good supply of bulblets for transplanting. You have to wait until all the flowers have withered, and then you jerk and twist the plant to get it into a position to layer. It is wise to keep your foot on the ground just next to the plant to keep the bulb from being pulled up when you do this. Then prepare good sandy loam or sand and peat moss to spread under the length of the stem, and cover about one third of the stem. With late-flowering bulbous plants, the bulb and its bent-over stem and leaves may be layered in a greenhouse or cold frame. (See also Bulbs in chapter 4.)

You can layer house plants by using two pots, but the most amenable method for them is air layering. Here you proceed with the steps for air layering as described earlier. You will probably have especially good luck with house plants because you can maintain a warm, moist atmosphere for the plants without the cold winds, soaking rains, hot days, and drying gusts you get with outdoor weather. Use damp sphagnum moss, as suggested earlier, but you don't have to be so careful in tying on the polyethylene because when the plants are indoors, they obviously are not subject to torrents of rain running down their stems to get into the moss.

House Plants

The most probable aim you'll have in mind when air layering house plants is to recondition and revive tall, old plants that have become leggy, lost their lower leaves, or just have outgrown their position in your rooms. If the plant is tall and you want to shorten it, put the slit and moss about 18 inches from the top. Or just use your judgment about where the new base of the plant should be. Then slit and wrap the stem at that spot.

Plants to try include Australian umbrella tree (*Schefflera actinophylla*), the coffee tree (*Coffea arabica*), coral berry (*Ardisia crenulata*), croton (*Codiaeum variegatum pictum*), *Dracaena* spp., dumbcane (*Dieffenbachia amoena* and *D. picta*), figs of various sorts, including rubber plant (*Ficus* spp.), philodendrons—especially the large ones—and Ti plant (*Cordyline terminalis*). The best stage is when the stem is half-ripe, and not yet coarse and woody. You'll be delighted with the stocky, handsome new plant you can get for yourself from this simple method of air layering.

Chapter 4

Division

The method of plant propagation called *division* is one of the easiest and most reliable methods there is. Each essential growing part of the plant is included, and you depend for success upon the ability of the plant to be severed and still survive. As with all asexual methods of propagation, division will preserve the purity of the clone (not permitting genetic variations such as those you may get when you propagate with seeds).

In simple division you merely cut through a clump of perennials or shrubs, or through a single bulb. This succeeds because you include both top growth and roots, or, in the case of the bulb, both a slice of the upper bulb and a piece of the basal plate.

When simple division is used on clumps of herbaceous plants or shrubs, it is also called *crown division*. Some kinds of crown division depend on common types of plant proliferation. A plant like a primrose, for instance, sends up new little plants around the mother plant. (These are often called *offsets*.) You merely separate the new plants from the old one by pulling or cutting them apart, and set the divisions loose to live independent lives. Bulblets and cormels taken from

Simple Division and Crown Division

larger, older bulbs and corms are divided the same way. In all these instances the result is an invigoration of the plants and more prolific bloom.

Another common kind of plant proliferation is by stolons or runners as described in the preceding chapter. These send out underground (or sometimes aboveground) shoots, which usually take root at certain points where they then create new plants. In propagating these stoloniferous plants, all you do is *cut the stolon* or long-running shoot near to the places where they have taken root and new plants are beginning to grow. These you transplant to another part of the garden. You can, for example, get a dozen or so new plants with very little trouble from a ground cover such as bugleweed (*Ajuga reptans*). In fact, many of the most useful ground covers are stoloniferous because such plants fill in space by themselves, without the bother of dividing and replanting.

As you might guess, there are some plants which definitely should not be subjected to any form of simple or crown division. These are the plants with taproots and the legumes like lupines.

When to Divide Herbaceous Perennials

Many herbaceous perennials die down by themselves in the fall, and that is a good season to dig up the plants very carefully to divide them. You can cut down through the top growth and the roots to make 2, 3, or sometimes 4 smaller clumps which you then transplant. Another method to use when the lines of proliferation are evident is to cut or pull apart the plant along these lines. The advantage of dividing perennials in the fall is that they can get established before the next season and will then not be likely to have a setback, which can cause reduction or loss of bloom for a year. The disadvantages are that the looseness of the soil around the transplanted perennial may cause heaving during winter and spring alternations of thawing and freezing, and may even cause freezing of the roots when the weather is severely cold. If the roots have been weakened or damaged, the newly divided plants may suffer rather badly. So be sure to provide a good cover of mulch to hold down the soil and protect the roots and crowns. There are, however, quite a few very vigorous plants which are customarily divided in late summer, for their capacity to grow good supporting roots during the fall season is perfectly adequate to the stress of being disturbed. Also, in fall, some plants about to go dormant react favorably to disturbance at that time.

Phlox is a plant which reacts much better, I have found, if divided in late summer or fall, though I have known gardeners who alternate dividing their phlox every 2 years in the fall and then every 2 years in the spring. Peonies definitely should be divided in the fall because you certainly would not want to disturb their tender pink sprouts when they are coming up in early spring. I think primroses (*Primula* spp.) respond well to fall division.

Many plants, however, really do better if divided early in the spring because they then have a whole growing season to adapt to new conditions. The rising of the sap helps to make them strong in spring, and the increasing length of the days gives them a chance for good healthy photosynthesis and the making of sugars. See that your divided sections are well watered when you replant them, and keep them shaded, too, if you see signs of wilting. Some booster fertilizer will help them to get reestablished more easily.

Plants definitely needing spring division include salvia, scabiosa, and spiderwort; and plants doing well after spring division include perennial ageratum, the artemisias, astilbes, bellflowers, bugbane, coreopsis, coral-bells, foxglove, Jacob's ladder (after blooming), potentilla, and veronica. As you do with other crown plants, be sure to save at least one piece of good crown for each section of roots you cut or pull away from the mother plant.

Techniques for Herbaceous Perennials

Crown division of phloxes and peonies in the fall can be done by digging up the plant and then putting 2 forks down into the crown, back to back, and simply forcing the clump apart. You can divide it into 2, 3, or 4 pieces, depending on the size. Be careful not to injure the fleshy roots of peonies, for that would cause a setback. With phloxes, remove the suckers (and any seedlings) which may have formed around the central part of the crown. If allowed to grow, these will produce magenta blossoms, and soon will turn the whole clump to magenta-blooming phloxes as they take over and dominate.

Primroses require a division that is somewhat different. With these you work according to the rosette structure of the plants. The central rosette has small rosettes which come as offsets around the mother plant. Each of the little ones can be severed with a knife or just pulled away from the old plant and transplanted to a cool, rather moist but well-drained place. All primulas need division every year or so, or the rosettes will get smaller and smaller and finally give up blooming. But if you propagate them, you can enjoy many, many new plants over the years. It will do no harm to lift primroses any time after they have finished blooming.

WARNING: With rosette plants the crown (where roots and stems merge) must not be put below ground level or the plant will rot. Clump plants, however, are not injured by this practice. In fact, some propagators habitually put many such plants with the crown ½- to ⅓-inch below ground level. With plants which strike roots very easily, such as tomatoes, it is helpful to plant even deeper than that because the many new roots will strengthen the growth of the plant.

Asters also need regular, frequent division. We have grown the ones called Michaelmas daisies, and found them very satisfactory and very easy to divide. Since they bloom in the fall, they can be divided after blooming or very early in the spring, though spring division means some diminishing of bloom that year. Use 2 forks, back to back, as with phlox, to force the clumps apart after lifting. The clump may be quite tough so force the forks way down in. Then separate the clump into as many pieces as will make nice new clumps. The central part is old and tired and must be discarded. You can see where the new clumps are forming so you can easily use those lines as marks for dividing the asters. You don't have to divide each and every one, but you might as well because they'll all grow old and multiply and need division within a few years. They make such a fine display in autumn that you will be glad to have a goodly number. And you can always give away extras.

Daylilies are less urgent, but it is better not to let them go many years without dividing. They need division when they get crowded, have many thick leaves, and begin to give up blooming. (The decline is not so likely with *Hemerocallis fulva*, the common roadside daylily which spreads by stolons and pretty well takes care of itself.) Use the same method of 2 forks, unless the clump has become so tough you need an ax.

One of the pleasures of dividing time for daylilies is that you have a chance to take off some of the delicious little tubers to boil up for supper. If you cut back the plants, you can also have a second crop of new young shoots to cut and eat steamed as you do asparagus. We have enjoyed these when we divide our daylilies in August.

Crown division is an especially good method to use for plants which grow near or in the water. This is because the shock of disturbance is minimized when the roots are continuously well watered and well accustomed to being in the water. Sweet flag (*Acorus calamus*) is very easy to propagate. It increases itself quite readily, of course, but if you want to divide it and move it around to other moist places, do so without worry. You can also easily divide the beautiful Japanese irises, which like wet places until they are through flowering. The best months to divide them are August and early September. The native iris of the South (*Iris fulva*) is an easy plant to divide, too. I have divided Siberian irises both in

Crown division of a daylily is shown here (1) using 2 forks to pull apart the sections of the clump after it has been dug up. Remove the leaves with a sharp knife (2) to reduce the leaf area and to help the plant adjust after replanting. (3) Replant each section of the divided crown in a large, well-prepared hole. A few tubers (4), or even one, if there is a shoot attached to it, will produce a new plant.

Water Plants Especially Successful

the spring and the fall, though their bloom is more continuous if you divide in fall. Otherwise it takes a year or so for the plants to recover. Good small sections help them to multiply rapidly, though larger sections give a good clump effect in a year or two. WARNING: Do not plant Siberian irises where water will stand over their crowns during the winter.

Other pondside plants respond to even simpler methods. Water-arum (*Calla palustris*) can be propagated in the spring with no setback. Just get down to the rootstalk and cut it with a sharp knife. If sections of the stem get broken off in the process, merely root them in some nearby mud. Both varieties of forget-me-not (*Myosotis scorpioides* and *M. sylvatica*) self-seed so rapidly all you have to do is pull up an old plant which has gone to seed and throw it down where you want new plants to grow. The beautiful lobelias, both red and blue (*Lobelia cardinalis* and *L. siphilitica*) can be propagated by simple division, by seeds, or cuttings taken from offshoots from the old stems. Vegetative propagation of some kind is advisable, especially for cardinal flowers, for they may be rather short-lived if you grow them in your garden.

Propagating Other Wild Flowers by Crown Division

Since I am a great believer in propagating wild flowers to help prevent their extinction, it seems fitting to recommend a few other plants you can bring in and propagate quite easily. A good wetland plant to try is the marsh marigold (*Caltha palustris*). This can be propagated the very first thing in the spring, or after the plant has flowered a little later. We know a place where this marigold grows alongside a ditch by the road, and though we do not do it anymore, we used to dig up a few plants early in the spring and bring them in to bloom in the house in a deep dish of water. Then, after flowering, it was an easy matter to divide the plants and put them down by our running brook. I think they really like more sluggish water; so I learned that the right thing to have done was to have put them back where I got them. And this is often the rule, I now believe, for most wild plants: put them back where you found them.

I have other reasons for wishing I had put back the fireweed (*Epilobium angustifolium*) which I moved. There was one clump to begin with, which I divided and then transplanted so there was one clump at the upper end of the field and another halfway down. But within a few years I had the whole place filled up with offspring. It was a beautiful pink sight in the summer, but enough is enough.

Loosestrife (*Lysimachia* spp.) is another flower that can take right over. If you want it, ask a good friend for a divison, and you are on your way. The fringed variety, *L. ciliata*, is a handsome plant with a long season of bloom and nice yellow flowers. Divide it at any time you wish.

(For other suggestions of wild flowers appropriate for crown division, see Part II.)

Plants needing acid soil (pH of 4.5 to 5.5) must be handled very carefully. They have delicate roots, or roots with no root hairs at all. (Examples are blueberries, azaleas, laurels, and most rhododendrons.) Therefore these plants must rely on certain bacteria and fungi, so any transplanted divisions must be moved to soil with the correct population of microorganisms. Some of the necessary fungi also need air, which accounts for the very shallow root systems on the heath family of plants. Be careful not to plant them too deeply.

The best place to put divisions from acid-loving plants is in a spot where other plants of the same species are already growing successfully. Azaleas especially prefer this, and also the andromedas called leucothoe. They might succeed if you move them into the shade of tall oaks, honeylocusts, laburnums, or birches, but never divide and move azaleas to places near shallow-rooted trees. And of course never plant any acid-loving shrub or plant near a cement wall, concrete walk, or limestone ledge.

Use only very light mulch or compost over the root area of acid-loving plants.

Crown Division of Acid-Loving Plants

Most herbs are dry-soil plants, but even these can be successfully divided if you work rapidly, on a cloudy day and if you have plenty of water on hand for a good soaking of the roots after you have replanted the divisions.

There are two or three ways of proceeding. Cut some with a

Propagating Herbs by Crown Division

French tarragon is one of the herbs that can be propagated by crown division. Dig up the clump (1) and cut it apart, making sure there is a shoot attached to each root part (2). A good dividing job will give you several smaller tarragon plants. (3).

sharp knife after digging the clump, doing as little damage to their roots as possible and being careful to separate the roots before replanting. Herbs needing this treatment include oregano, tansy, hyssop, mugwort, salad burnet, southernwood, violets, and valerian.

Members of the onion group (*Allium* spp.), including chives and some of the ornamental alliums, can be dug up and just forked apart. Others to fork apart, but very gently indeed when you get down to the roots, are sweet woodruff, German chamomile (*Matricaria chamomilla*), bee balm, boneset, and bugbane. Mints are less tender, and have, as I mentioned in the chapter on layering, such tough stolons that you can practically chop through the plants anywhere and get pieces to transplant.

More delicate herbs with very tender roots, such as sweet cicely (*Myrrhis odorata*), must be handled extremely carefully or not divided at all. I'd really recommend making cuttings instead, as I would for wintergreen (*Gaultheria procumbens*) also.

Root Division and Root Pruning

When the plant you want to divide is very tender, or has a taproot, root division is a safer method than crown division. Root divisions are like root cuttings, and depend on the ability of both fleshy and wiry roots to send up new shoots when pieces are severed from the mother root. Not all plants are amenable, some do not react so readily to the stimulus of the wound. And whenever you work with roots, of course, you must protect them from drying out from exposure to the air. With this method you can lift the plants that do not have taproots, and simply cut off pieces of root to replant. Do this in damp weather, or an hour or two after a good thorough watering of the plants. Choose good, stout, healthy roots— usually side roots—and cut them with a sharp knife so that you get several pieces suitable to use for replanting.

With taprooted plants like oriental poppies which do not wish to be disturbed, never attempt to lift the whole plant. Just move the soil and feel down to get to some of the side roots which you can cut for propagation. This method is a good safe one, in fact, for all plants. The tuberous roots of sweet potatoes, dahlias, and tuberous-rooted begonias are divided by cutting off the section of root with a growing point, which is that nearest the main plant. Then you plant the section with this point uppermost.

Be sure to plant all root pieces right side up, rather shallowly so that they will have access to oxygen and the encouragement from light they need for shoots to grow as soon as they are initiated. To keep from confusing root tops and bottoms when replanting, you can cut the top end straight across and the lower end on a diagonal.

Root division is a good method to use on delicate plants which do not really like to be disturbed. I recommend it for both wild flowers and herbs that do not stand the shock of moving very well. See Part II for examples. Some herbs such as chicory and endive, are often divided and brought into the cellar to start growing in late winter for greens to eat. Others like lovage and sorrel react better to root division in the spring.

Root pruning is very much like root division. The roots of shrubs and trees often are pruned, however, for reasons other than propagation. One reason to dig down and cut back some of the roots is to prepare a shrub or tree for transplanting. By doing this you shorten the long roots, encourage growth of fibrous, tough side roots, and strengthen the plant before the shock of moving it. A second reason to cut back roots is to remove a girdling root, which should be eliminated lest it choke the tree.

Root pruning preceding the crown division of shrubs will help them to adjust to the disturbance. The way to do it is to take a sharp spade and cut straight down through the side roots about 16 to 18 inches from the main stalk on small shrubs, and farther out on larger plants. Then leave the plant in the ground to grow new roots for at least 6 weeks (and even up to a year) before dividing it. The pieces of root you cut off can be used for propagation, of course, if you wish to.

Division of Bulbs and Other Underground Stems

Propagating bulbs and other underground stems, corms, and tubers involves both the original and yearly cultural practices and methods you use to increase your supply of plants and blossoms. Sometimes it is merely a process of dividing the bulblets from the mother bulb, offsets from the mother plant, or the cormels from the mother corm. (See simple division in the beginning of this chapter.) Sometimes, as with hyacinths, you scoop the bottom of the big bulb to wound it, and wait for new bulblets to come. Plants having rhizomes—for example, iris—are divided so that each section of rhizome also has a section of the upper plant attached to it. Tubers are usually gently cut away from the clump either before or after winter storage, and then planted separately. Stolons and runners are merely severed, and the new plants moved as desired. Hardy bulbs such as narcissi, tulips, and some lilies can usually be left in the ground over the winter without being injured by frost (or put in cold storage). Other tender bulbs or corms must be lifted, divided, and stored in a dry place that is consistently warm enough so they never suffer from the cold. As for all methods you use for propagating, your utensils and containers must be very clean. And you should be on the lookout for diseases and infestations so you can burn up what is infected.

Bulbs

Perhaps the most common method of propagating bulbs is to lift them after their green leaves have died down, and divide the bulbs by just separating them, or by removing the small bulblets which have formed. This is usually done in early summer, as with the various varieties of narcissi; but tulips must be lifted later because you don't want them to be spurred to grow during warm spells in autumn, after which their tender leaves might be nipped. Both are lifted in summery weather, however, so that their roots will have a chance to grow before the ground freezes hard.

You can judge that it's the proper year to lift and divide bulbs of narcissi and tulips when you notice that they tend to grow more green leaves than shoots which blossom. Narcissi leaves will get very thick and luxuriant; tulips will send up a single, large green leaf without any bloom at all. The time to divide varies, but it may be within 2 or 3 years of the time of planting, though every 4 years is often enough for most narcissi. Various members of the lily family, such as the fritillaria characteristic of formal gardens, should also be lifted every 2 or 3 years. Naturalizing narcissi can be left where they are indefinitely.

Drying It is best to let the narcissus bulbs dry in the air in part or full shade for a few days. If there are plenty of little bulblets to take off, you can plant them in a nursery bed until they are large enough to bloom. If they are planted 3 inches apart, under 4 inches of soil, they are usually large enough to move in 1 to 2 years. I often plant them right back in the daffodil bed, and let them develop right there until they are ready to bloom. Tulip bulbs may also be dried off. If you handle them this way, you can lift them earlier in the summer, and then store them until it is late enough for planting. People who rotate large plantings of tulips for display often handle their propagation and culture of tulips this way.

Slicing It is possible to propagate some bulbs by making careful slices with a good sharp knife right down through the bulb from top to bottom between the scales, as long as you see to it that each slice has some of the basal plate included. With small Dutch irises or small scilla bulbs, for instance, you can make 8 cuts so that with each you expose the scale rings to guide you for the next cut. Slip your knife down between every 3 or 4 pairs of scales, and on down through the basal plate, which must be included for the rejuvenation and

(1) A bulb can be divided by slicing straight through it with a sharp knife so that each slice includes some of the basal plate. (2) In the sliced bulb here the scale rings are exposed. Eight such slices can be made, even in small bulbs.

propagation you want. The pieces then can be handled as though they were leaf cuttings, and planted in a starting medium until they have grown enough to transplant. This is a good method to use on half-hardy bulbs like amaryllis (*Hippeastrum* spp.), Aztec-lilies (*Sprekelia formoisissima*), rain-lilies (*Cooperia* spp.), or Guernsey-lilies (*Nerine* spp.) as well as daffodils, if you want to vary your method of simple separation. Small bulbs like scillas can be handled so, too.

Scooping Scillas may also be scooped, but you are more likely to do this to hyacinths. It is actually a technique for encouraging new bulblets and is used for hyacinths because they normally do not send out those new little growths themselves. You make wounds which then induce the bulblet growth. First lift the mature bulb, then take a melon-baller gadget from the kitchen and just scoop out the bottom of the bulb. You can use a steel teaspoon instead. Cut in far enough so that you destroy the main shoot, going right through the center of the basal plate. A variation on this method is to cut 3

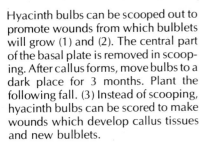

Hyacinth bulbs can be scooped out to promote wounds from which bulblets will grow (1) and (2). The central part of the basal plate is removed in scooping. After callus forms, move bulbs to a dark place for 3 months. Plant the following fall. (3) Instead of scooping, hyacinth bulbs can be scored to make wounds which develop callus tissues and new bulblets.

crosscuts across the base of the bulb, again going right through the basal plate. Make the cuts so they meet in the center and create a 6-pointed star.

Do not, of course, use any bulbs that look diseased, and dip your spoon or scoop or knife in alcohol or some other disinfectant between cuttings. It is wise to dust the cut bulbs with sulfur before you put them on open trays or in dry sand for about a week, or for whatever length of time it takes for callus to form at the sites of the wounds (at the axils of bulb scales). Keep the treated bulbs at about 70°F. for best results. Then when you see that callus has formed, move the bulbs to a somewhat dark place for about 3 months. The darkish place can be as warm as 85°F., but increase the temperature to that degree very gradually. Take about 2 weeks to do it.

In the fall, plant the hyacinth bulb with its bulblets 4 inches deep and the next year the bulblets will send up many green leaves. The mother bulb will disintegrate, so dig up the group the next fall, separate off all the bulblets and replant them to grow and develop to flowering size. Treat small bulbs the same way, but plant them at depths appropriate for their sizes.

Usually a scooped bulb will come to flowering stage in 3 or 4 years. Hyacinths will produce up to 24 bulblets, so the method is well worth it. You can consider the bulblets ready to plant in the garden beds when they get to be 5½ to 6 inches in diameter, but for forcing indoors or in a greenhouse they should be ½-inch larger.

Small bulbs take less time, but need both the resting time and the warm time in a damper, darkish place. Of course the tools you use on small bulbs need to be smaller. The kind of scalpel used in biology labs is good and strong and easy to handle.

Bulbils are small new growths which come at the axils of leaves of some lilies, for instance the one we call the tiger lily (*Lilium tigrinum*) and also (*L. bulbiferum*). They look like rather large black seeds, and often grow up into new plants when they just drop to the ground. But you can collect them, too, and start them in a nursery bed. You'll get a great many from each lily, and if you pinch off the flowering top you'll probably get even more. This is undoubtedly one of the easiest methods of propagation there is.

Some lilies also have little new growths like bulblets at the base of their stems. These you remove in the early fall, sow about 2 inches deep and 2 inches apart in a nursery bed and leave to grow for a year.

Scales, which are modified leaves, are conspicuous at the bottom of the stalk of some lilies. You dig around the larger bulbs soon after the lily has flowered to find them. Carefully take off a few of them, without injuring the roots at all, or

Removing Bulbils and Scales

digging them up. Some people also remove a few scales from new lily bulbs they have just bought, but you must not take off so many that you disturb the strength of the bulb. You can dust the scales with rotenone and then store them in a polyethylene bag. It helps to put a little slightly moist vermiculite in the bag to keep up a minimum of moisture. Tie the bag tightly, and put it in a warm, sunless place that has some light. You will see the very small bulblets begin to appear within 2 or 3 weeks. Keep watch until the rootlets are ½ inch long, and then you can plant them when autumn comes.

Another way to propagate by scales is to lift the lily bulb, for instance, and carefully remove them. The time to do this is as soon as feasible in the autumn after the lily plant has completely died down. Plant them several inches apart in a shallow furrow, in a well-drained, somewhat shady place. Line the furrow with sand, and you can add some sulfur to be on the safe side. Cover the scales with up to 2 inches of sand, or, if you prefer, peat moss. Watch for the little bulblets which grow along the edge of the scales, and when they crowd together, lift the scales from the sand, separate the little bulbs and plant them in the nursery garden or cold frame.

Rhizomes   The fleshy, horizontal underground stems of plants like irises and quack grass are called *rhizomes*. Some plants with rhizomes are greatly helped by lifting and dividing, but others like lilies-of-the-valley just go on and on without being disturbed and bloom very well. Irises, however, must be divided to keep them flourishing.

Here are iris rhizomes showing roots and new shoots (1). To divide, sever them with a sharp knife so that each piece has roots, a shoot, and a fan of leaves (2).

80

Lift the whole plant in middle or late summer and cut the rhizome with a sharp knife so that each section has a good part of the top plant as well as of the fleshy rhizome. Be careful not to injure the roots. And do not wait until late fall, for then the roots will probably die for lack of food. When you replant the pieces, lay them just parallel with the surface of the earth, and with their tops exposed to the sun. Rhizome division can be used for propagation of plantain lily, canna, false Solomon's seal, mayapple, and trillium. If you propagate them when dormant and cut the rhizomes near the nodes, they ought to do well.

Ferns can also thrive on rhizome and clump division. If you have some in your garden already, it may appeal to you to divide them and get a lot more. Even if you have to buy your first specimens (or have them given to you by a friend in the woods or on the byways), you can increase your supply quite easily.

The rhizomes and masses of roots of many ferns are very fibrous and tough, so provide yourself with a big, sharp blade or spade to slash through the matted growth. Dig up the fern, then make a good thrusting cut right down through on the line of division you have selected. Prepare the new site for the fern beforehand, and then move the pieces promptly to the spot where the divisions are to grow. Take a good big root ball with each piece, and introduce some of the fern's familiar soil into the newly prepared ground. If the plant has grown where there are stones and pebbles, be sure to put in some of those, too. Plenty of leaf mold or humusy soil, and a good mulch, will also help the moved clumps or rhizomes to survive. Water well, and keep moist, especially in dry, hot weather.

The ferns which have creeping rhizomes are easier to divide. Also easy to divide are the little plants which are sent up at the nodes on runners of some big ferns. Sever the runner, and if the weather is sunny and warm, leave the plantlet where it is for a few days to adapt to the separation. (Provide shade if needed.) Or at least wait until a cloudy, cool day to move it. When you do move plantlets, or divide rhizomes, water well to help them adjust.

Indoor ferns which have rhizomes hanging over the edge of

the pot are the easiest of all to divide because you can layer the overhang first, putting it on or into a moist pot of fern-soil, and then just wait for the results before you cut it off. Other ferns grow many plantlets on their pinnules, or subleaflets, which you can remove when they have three or four leaves.

(See Spore-Bearing Plants in chapter 6 for directions on propagating ferns from spores.)

Tubers

Sometimes a rhizome has an enlarged, swollen section at the end. This is called a *tuber*. Familiar examples are the tubers of dahlias and potatoes. You must divide the clusters of dahlia tubers yearly, being careful to keep some of the upper plant, or crown, attached to each division. Potato tubers, as you probably know, may be entirely separated from the

Each dahlia plant develops a half-dozen tubers during a growing season (1). Divide the tubers with a sharp knife, keeping some of the stem with each division (2). Then the separate tubers (3) are ready for spring planting after the soil warms up.

mother plant and either planted whole, or cut so each piece has at least one eye. This eye consists of a very small ridge which has on it a tiny, scalelike leaf. In the axil of this leaf are three or more tiny buds. If you examine one of your sprouting potatoes sometime, you can probably see these little structures. The pieces (or whole potato) may be planted several inches deep in the earth, or placed on top of the soil in deep mulch.

Tubers are rather tender, and in order to maintain good stock for propagation, you should prevent their getting frozen, split, injured, or soggy. And never injure the root with the growing tip; it's the one nearest the mother plant. Get tubers good and dry before storing by leaving them in the sun for a while. Store your tubers at about 45° to 50°F. In preparing them for planting, always use sharp, clean knives. They can be planted as early as February if kept in warm, damp peat.

Another kind of underground stem is the *corm*. This structure is rather round, and bulblike in appearance. Like aboveground stems the corm has nodes and internodes, but like bulbs and rhizomes it is a food-storing part of the plant. (In contrast, bulbs have scales wrapped around a very short stem.) During the summer the old corm forms little cormels around the edge of the basal plate. When the foliage dies down on a corm plant like gladiolus, about 6 weeks after blooming, dig for the corms. Cut off the leaf stalks to within 2 inches of the ground and lift the corm itself very carefully. Put

in flats to dry in an airy place, and when well dried, take off the cormels and any new large corm that may have formed. You can discard the old corm and its roots. Store at 40° to 45°F. until warm weather the next spring. The large corm will probably bloom the following year. The cormels take 2 or 3 years to bloom.

One exception to spring planting is the autumn crocus, which is planted in August. These crocuses bloom in fall, and send up green leaves in the spring. After these leaves die down you can lift the corms and divide them. You only need to do this at 3- or 4-year intervals, because they are hardy.

When you dig for corms, feel around with your finger in the earth to see whether there are any extra little corms on runners besides the main ones on the central plant.

A second method for propagating corm plants is to cut the main corm into several pieces. If you do this, you must include a part of the basal plate in each cut so that the new plant will have the right spur to grow. A good way to strengthen the spur is to let the cut surface dry and callus. It

Divide the corms of gladiolus by gently pulling them apart before storing for the winter (1). Remove the cormels growing from the corms (2) and plant them in the spring in a nursery bed for maturing. Once mature they can be transplanted to a permanent location.

83

may help later to soak corms susceptible to disease in a gallon of water with 1½ tablespoons of Lysol® in it. The best time to do this is just before planting.

Other Methods for Plants with Underground Stems, Bulbs, Etc.

Leaf cuttings can be made from various tender bulb plants, such as the gloxinias. The new lily plants that grow from bulbils can be bent over and layered. And many lilies and other plants discussed in this chapter can be planted from seeds. Cyclamens, various lilies, gloxinias, and the small bulbous irises self-sow quite readily. I have also had small alliums propagate themselves by self-sowing, as well as by bulblets. And perhaps you know from experience that the star-of-Bethlehem, a small bulb plant, will propagate itself all over your garden if it once gets started.

Chapter 5
Budding and Grafting

In budding and grafting you propagate a new plant by bringing together growing parts of two usually separate plants to make a combination unlike any that nature ever invented. When the parts brought into union are relatively mature and from relatively large plants, the term *grafting* is used. *Budding* is an increasingly common kind of grafting involving young plant tissues: the *rootstock* or *stock* (the lower part of the new plant) is often a seedling, and the *scion* (upper part of the new plant) is a single bud, in contrast to the two or more buds on a grafting bud stick. The stock develops into the new plant's root system; the scion into its branches and leaves. The stock can also constitute the new plant's lower trunk.

There are many good reasons for grafting, and it is not too hard to learn some of the techniques, even if others are rather delicate and difficult to master. Over the centuries gardeners who have mastered the art have been well repaid for all their exertions when they have created plants of high quality and fruits with improved taste, shape, color, pest resistance, and other superior characteristics. But there are other aims in grafting, too.

Sometimes a tree injured in a storm needs to be repaired. Sometimes people want to change a bearing fruit tree, for instance, into a new variety. Sometimes a house plant fancier wants to change a short-day plant like *Kalanchoe blossfel-diana* into a long-day plant that will bloom in a season other than winter: this can be done by grafting it onto a showy sedum (*Sedum spectabile*), a long-day plant that will influence the kalanchoe to change its ways. Professional propagators use grafting to get excellence in woody plants, putting rare and beautiful lilacs on privet stock, for instance, or fine crab apple hybrid scions on more rugged and disease-resistant stock, or unusual tender roses on hardy, cold-resistant stock. Though grafting does not result in obvious rejuvenation as other propagation methods do, it is just as exciting, for it makes possible fruit trees, nut trees, ornamentals, and even new herbaceous plants which you simply cannot have in any other way.

What Happens in Budding and Grafting

Besides dwarfing, the big obvious changes budding and grafting can bring about include earlier bearing and increased hardiness; the preservation of juvenile characteristics; and strengthening or invigoration, as when roses were grafted onto multiflora roses, or when an unpopular tree or shrub is transformed into a popular one, whether it be an ornamental or a fruit or nut tree. (Sometimes you have more trouble with suckering from the vigorous rootstock—as with plums under other fruits and privet under lilacs—but the vigor is worth it, and suckers can always be cut back.)

The fact that good, successful grafting can produce so many changes in the habits of plants somewhat distracts us from the universal, underlying conditions and events in plant life on which these changes depend. There are several basic plant processes which play a role in grafting.

The Wound and the Healing of the Wound

Though propagators aim for similarity between stock and scion, there are one or two big differences between these two kinds of plant material. The stock that is chosen is really older, though the area used for inserting the scion piece is from a part of the plant still characterized as young, that is, it is capable of meristematic growth and is still making fresh, young undifferentiated tissue.

When this young, active tissue is cut and brought into close contact with the freshly cut tissue of the scion, the grafter fits the two together so that the cambiums meet as well as the propagator can possibly make them. As I explained in chapter 2, botanically speaking the vascular cambium is a lateral meristem which produces new phloem cells toward the outside and new xylem cells toward the inside. In general garden terms, however, the cambium is thought of as including all three: the vascular cambium, the phloem, and the newest layer of xylem, but especially the last two. So when a grafter takes care to match up what he calls two cambiums, his aim is to take care of the conducting tissues which make it possible for the newly grafted plant creation to survive. But whether the grafter is aware of it or not, he also takes care of the very center of life within the plant: the very sensitive and essential cambium layer whose one job is to make new cells.

It is the wounds inflicted by the grafter when the stock and scion are cut before they are matched which stimulate cambiums to special action. Instead of continuing to proliferate phloem and xylem cells, they begin to make callus, that is, tender, thin-walled parenchyma cells which immediately begin to fill in the minute space between stock and scion when the two are placed together. Soon the two new sets of callus meet and their cells interlock. This is the important moment. From then on these interlocking cells are the cause of what happens.

The parenchyma callus begins to produce or differentiate into new cambium cells. At last these link up with the cambiums of stock and scion so there is a newly linked layer running right up and down the grafted plant. And this can then produce new vascular tissues, so that the phloem and xylem will be healed and lined up to carry the water, mineral nutrients, and plant sugars up and down the healed system. Roots grow, new leaves come out, the tendency to wilt passes by, and the graft can be said to have taken. With budding grafts it is at this time that propagators begin to cut back the top of the stock above the graft, for it is no longer needed and might rot or shrivel if not removed.

Luckily this same series of events will take place if there are

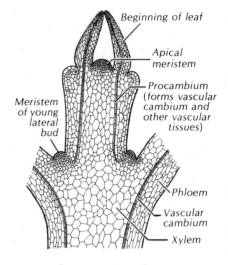

Cross section of a growing stem.

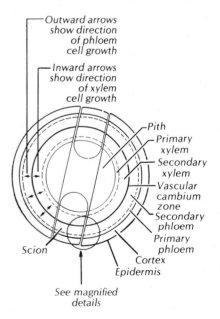

Diagrammatic cross section of a stock with two cleft graft sections.

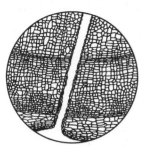

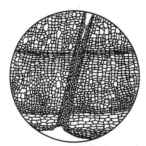

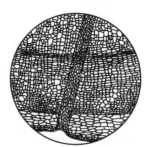

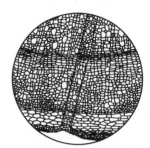

Step one—*production of callus tissue in cambium region*

Step two—*interlocking of undifferentiated tissues called meristematic growth*

Step three — *differentiation of cells into new cambium*

Step four—*production of new vascular tissues by the new cambium, permitting passage of nutrients and water between stock and scion*

Developmental sequence during the healing of a graft union

three pieces involved, not just two, as long as they are all carefully matched. (See later section on Double-Topworking.)

It is interesting to know what happens to the cells that were actually wounded during the grafting process. They die and form a necrotic plate of material that is reabsorbed later. If they happen to include the vascular tissue cells, those are gummed over. And it is from under these wounded tissues that the parenchyma cells begin to appear in a few days or so.

One of the advantages of using older stocks, or at least stocks that are older than the scions, is that older plants are able to produce more parenchyma. Evidently the cambium of the stock is most active in forming the new cells; but evidently the new scion tissues become resonsible for the new activity once things get lined up. All this must happen before new leaves come, or the scion will die of starvation from lack of phloem and new xylem tissue to provide for the new growth.

Once in a while there are exceptions to that sequence, and when there are, one kind of cell just takes over an unusual function until things get straightened out. Probably all of these wonderful happenings are influenced by auxins from the buds, and of course sugars made in the green cells provide the energy for the new growth.

Success in Budding and Grafting

1. Compatibility

Successful budding and grafting depend in the first place on compatibility of materials. There are instances where material from a plant is put back on itself. That obviously works. In other cases near-relatives are used; sometimes these are compatible, sometimes not. And propagators also match up

88

more distant relatives in combinations like quince rootstock and pear scions, which experience has taught are compatible. Compatibilities of this sort are very important. For example, many a fruit tree planting has been made possible and pleasing because the graft on quince works, producing a dwarf tree that fits conveniently in the yard and fruits earlier than trees of standard size bearing the same fruit.

Compatibility may also be linked to similarity in vigor, where the growing rate and strength of one plant resemble those of the other. It may also have to do with similar yearly starting dates for vegetative growth so that one plant is not zooming ahead of the other and in effect stunting it. Compatible plants need to have more or less the same physiological makeup, or the same sort of biochemical activity. Too much of one kind of acid in the stock, for instance, can injure the tissues of a scion which is not used to that chemical makeup. When circulation of nutrients from stock to scion introduces the alien substance, growth patterns may be disrupted, especially in the important and sensitive cambium.

2. Cambium Connection

For a successful graft, the vascular cambium of both plants must be matched as closely as possible. Though the materials which come from the calluses put out by each partner are the really significant link, the flow up and down of water, mineral nutrients, and sugars has to go through the phloem and xylem cells adjacent to the cambium. Because of this, their correct adjustment to each other is essential. They simply must line up or the graft cannot take. In fact, one main reason for putting on tape, or other strapping, is so that the phloem and new xylem cells will line up correctly.

3. Moisture and Oxygen

Plants that are grafted need both plenty of water and plenty of humidity to keep them going, to promote good callus formation, and to assure quick, proper healing. High humidity is important for the entire rootstock, and for the area where callusing is taking place, 100 percent humidity is quite proper. If the cells in the stock plant are turgid, with plenty of water in them, the proliferation of callus will proceed well.

The watery parenchyma cells of the callus are so very sensitive to drying that dry air will cut short their length of life. Sometimes tape is used to help keep the callus and other tissues from drying out, but in all the old-time grafting techniques and many of the new ones, too, wax is the means used to keep out drying winds and hot evaporating rays of the sun. (Waxing mixtures are described below.)

Soil moisture is also important to the seedlings or rooted cuttings used as stock. Since the plants are severely shocked by the wounds they receive, plenty of water is needed to keep the cambium active and its cells dividing.

Sometimes it protects and helps a graft to bury the point of union in the soil. This not only aids in maintaining moisture, but encourages good healing if conditions are favorable. Most important, it enables the scion to grow roots, discourages suckers from the stock plant, and establishes dominance of the scion. (See Part II for examples of fruit or nut tree grafts so helped, but note that some dwarfing stocks, especially those for apples, must not be put below ground level.)

The only exceptions to providing high moisture in the soil around the stock plants would occur in grafting plants with a tendency to bleed, such as maples, birches, and beeches in the spring.

Because of the rapid cell division and fairly high rate of respiration in grafted plants, oxygen is also necessary. If the plants being grafted have a marked need for it, it's best to omit wax and put them in a moist medium, as is the practice when grafting grapes. If watched to be sure they are getting enough oxygen, very small grafted plants can be put under plastic bags or jars.

4. Similar Physiological States

As we all know, plants go through various kinds of dormancies or near dormancies. It is not wise to try to graft together two plants in different states. If one is dormant, the other should be also. If one is in the first flush of spring growth, so should the other be. If one is full of sap and mellow at the end of the growing season, the best graft will come from a plant in like condition. Luckily, today we know much more than our ancestors did about the physiological details of these various states; we also have refrigerators which will stay at the convenient temperature of 40°F., if we want to slow down or immobilize a plant at a certain stage of growth.

The most important state for good grafts, of course, is good health. If there is any question at all about the condition of the two plants, any sign of virus disease or any other disease, do not use the material because that is just asking for trouble. Keep clean everything that you are working with; use wax and tape to keep out not only rain but bacteria and fungus spores, too. And if you smoke cigarettes, remember that your fingers can be carriers of the tobacco mosaic virus.

5. Good Growing Conditions

Grafts that succeed are those that are inspected and taken care of. They should not be knocked about by winds, machines, clumsy dogs and people, or rolling toys. They should not be choked by weeds or suckers from the stock plant. Even years later, when the graft is presumably well established, a stock plant will send up suckers that weaken the tree. Take care of them. Stake weak plants that need it. And if you live where rabbits and deer can get at the plants, put up wires; or put collars around young trees if smaller creatures like mice can come and nibble.

6. Right Equipment Rightly Used

Satisfactory results depend not only on the plants selected and conditions provided but also on suitable equipment properly used. Good knives are the first essential for good grafting. Any very sharp, very reliable knife will do, but it is best to get special folding knives. Called *grafting* or *budding knives*, these tools are strongly built and have blades of high-quality steel, set at an appropriate angle in relation to the handle. A good budding knife has an extra blade with a rounded end for lifting the flap on the T-cut. You also should have two good grindstones and should take a few lessons in the right way to sharpen a knife.

Waxes Materials for waxing are also important. A standard wax mixture to melt down and mix up includes 5 pounds of resin; ¾ pound of good beeswax; ½ pint of raw linseed oil;

Equipment for grafting is shown here: tape, wax, brushes, clippers, and a budding knife.

1½ ounces of fish glue, and 1 ounce of lamp black. Heat and combine all ingredients but the glue until they are well mixed. Use a double boiler to be on the safe side. When mixed, cool. Melt the glue with very little water in another double boiler, and when the other ingredients are somewhat cool but still liquid, pour in the hot glue, stirring all the time. Then cool the whole mixture and pour it out in shallow pans lined with waxed paper. After it hardens you can break it up into pieces and store.

When you are grafting, take with you a melter of some sort, perhaps an alcohol burner with a stand for holding a container for the wax. Melt the wax slowly and thoroughly, but do not let it heat up or bubble, for then it would be too hot for the delicate plant tissues to which it will be applied. Apply with a brush (which you should not leave in the hot wax between applications).

If you want to simplify matters, you can now buy grafting waxes made of asphalt and water, which can be applied cold. Apply these thickly, for the water evaporates to leave the coating material. A disadvantage is that such waxes may wash off if rain comes before the asphalt has had a chance to harden. By the way, never use a roofing mixture for waxing plants.

Another type not requiring heating is a hand wax made of a mixture of wax, resin, and tallow. Make it like taffy, pulling it until it has a nice consistency for putting on the plants and a pleasant yellow color. Proportions may be either 3 parts tallow to 1 part beeswax and 4 parts resin; or 1 part tallow to 2 parts beeswax to 4 parts resin. This blend is rather tedious to apply and suitable only for small grafting operations. To store it, roll it into balls and wrap.

Tapes You can tape two parts of a graft together with mere string, which you then cover with wax. Or you can use waxed string to begin with, preferably the kind which will stick to itself when you wrap it around the stem. Make it by soaking No. 18 knitting cotton for 10 minutes in a hot mixture of 2 parts beeswax, 1 part tallow, and 4 parts resin, to which you add 1 part linseed oil and ½ part paraffin. You can also make

You're going to have to melt your grafting wax mixture as you make the graft, so take along with you the necessary equipment, including a portable heater like an alcohol burner or a can of Sterno®.

waxed tape by soaking strips of cotton in this mixture. Adhesive tape is all right to use for whip grafts, and adhesive latex tape that nurseries use will do for all types of grafts. Cut the wrapping as soon as there are any signs that the stock or graft union is being strangled. One way to avoid the need for cutting the tape (if you are going away on a vacation, for example) is to bury the plant to a point above the tape and let the wrapping rot off in the moist soil. If you use plastic or polyvinyl chloride tape to bind a little bud graft, the end must be folded under because this tape is not adhesive. You can also use rubber bands.

Materials for Storing Scions   If you're doing spring budding or grafting, you can cut unfrozen scion wood during the winter and keep it dormant until your rootstock is growing well. You'll need barely moist peat moss or wood shavings to wrap around the scions and polyethylene plastic to bundle them in for storage in the refrigerator (at 40°F.) or in a cold frame until your rootstocks are ready. (For complete directions, see Spring Budding.)

Whitewash   Sometimes newly grafted plants exposed to the sun benefit from a white covering to prevent sunburn on their trunks, branches, and the new scions. If you need protection for a season, it is simple to mix up equal parts of water and water-base acrylic or latex indoor white house paint. You can also mix up an old-fashioned whitewash of 1 part sulfur, 2 parts salt and 10 parts quicklime. Carefully add some water to the lime to start slaking it, and then add the other ingredients. Use a crockery container, and make the whitewash a few days ahead. Before applying it, thin to a consistency proper for a spray or brush. For fall grafting, when the weather is cooler, you can easily cover the scion with a paper bag on occasional hot days to protect it from the sun.

Watering Equipment   You must have a good supply of water for graft union plants, especially when the graft is first made. The parenchyma cells which make up callus are more than 90 percent water, so obviously can't be formed without it. If you do not have a hose nearby, provide yourself with a good watering can. For difficult materials like evergreens you will also need humidity cases, misting arrangements, or plenty of polyethylene film to keep them moist.

Clippers If a lot of water sprouts spring up along the trunk or branches of the grafted tree, you'll want clippers to cut them back so they don't shade the graft union too much. (These volunteer shoots can rob and interfere with the newly grafted scion, but if the area of the union needs a little shade, you can leave on a few of the most useful.) The first year after doing cleft grafts or topworking, you will also need good sharp clippers to remove the unnecessary second-rate scions. They are useful, too, for getting rid of suckers.

Budding

In the old days most discussions of grafting started with the methods used in combining fairly large plants, but modern methods are so often based on budding very young plants that I would rather start here.

In budding, the stock, or rootstock, used as the underneath plant is a mere seedling of the variety wanted for hardiness, adaptability to certain soils, vigor, early fruiting, or other characteristics it may impart to the scion. The scion itself is a mere bud—usually with a little bit of stem or bark—which is slipped into a cut in the rootstock to make the graft union. The matching of the stock and scion cambiums is usually easier with young, small plants than with larger, older ones.

Condition of Rootstocks

The diameter of the stock at ground line should be about that of a pencil. Though the stock may be a seedling or a rooted cutting, the root system on a seedling is usually stronger and fuller. Layered cuttings are also sometimes used—rhododendron layers for instance—and they, too, often have good root systems. If the seedling you use is of a fast-growing species, one season's growth may be sufficient. (Then you do the budding in the fall, preferably when the temperature is about 80°F., but not as high as 90°F. and certainly not below 45°F.)

It is also essential that the rootstock plant be healthy and vigorous. Weak stock produces the effect of dwarfing. The stock should also be free from disease. In fact, one advantage of using seedlings instead of cuttings or layerings is the lessened likelihood of carrying over a virus disease from the mother plant. Also, for T-budding and most variations, the bark of the stock must be in a condition to slip easily, as it is during the active growing season. If the budding is done too late in the fall, for example, the bark will adhere to the wood because active growth has stopped. Then it will be impossible to raise the bark and make the T-cut without ruining the tissues. While the bark is slippery, there is active cambial growth, stimulated by the expanding buds on the plant and by the auxins and gibberellins they send out. It is during this phase, while cell division is still going on, that you should make most bud grafts. Even when small new shoots of older trees are used for the stock, if single-bud grafts are inserted, the condition of the rootstock should be just about the same, from the size of the stem to the active growing condition.

Like stock plants, the bud sticks from which you get buds for budding should be vigorous and disease-free. It is important to get buds from stems showing only vegetative growth, with no flower buds. I have seen recommendations simply to remove the flower buds when making cuttings and doing grafting, but it is always better to use a stem which has no flower buds on it at all. And with budding, when you use only one bud anyway, it is absolutely necessary if you want to make a decent new plant. You can tell the difference if you look carefully because the flower buds are almost always rounder and plumper than the buds which will bring forth green growth and no flowers.

Condition of Bud Sticks and Buds

Try to select buds from the middle or lower parts of a stem unless (as with cherries) those are the flower buds. Remove the leaves from the stick as soon as you cut it, and just leave a short piece of the petiole, or leaf stalk, to use as a handle. Keep the severed sticks cool and moist, putting them in moist newspapers or burlap if you aren't going to insert them immediately.

How to Cut Rootstocks and Bud Sticks

T-Budding

The cut you make on the rootstock will usually be T-shaped, which makes it fairly simple to slip the scion bud between the flaps at the top of the T and to slide it down the stem of the T until it is in place and ready for binding. After you make the T-cut on the stock plant, you should cut the bud in the form of a shield; in fact, this method is sometimes called *shield budding* instead of *T-budding*.

To prepare the scion, start the cut about ½ inch below the bud. Use a very sharp knife and slice upwards, as thinly as you can, until you get to about 1 inch above the bud. Then bring your knife out again. (When I say you should attempt to make the cut as thin as possible, I mean that you should try to

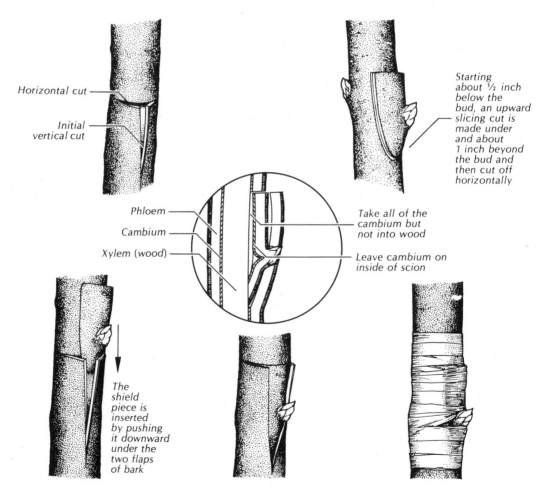

Horizontal cut

Initial vertical cut

Starting about ½ inch below the bud, an upward slicing cut is made under and about 1 inch beyond the bud and then cut off horizontally

Phloem

Cambium

Xylem (wood)

Take all of the cambium but not into wood

Leave cambium on inside of scion

The shield piece is inserted by pushing it downward under the two flaps of bark

T-Budding

avoid cutting into the wood. But by all means don't cut so shallowly that you cut into the cambium, which is just between the wood and bark. That would be disastrous, for an uninjured cambium contains exactly the materials you want in order to make the budding a success.) Next, detach the bud by making a second, horizontal cut at the point where you removed the knife. Then complete the shield shape, straight at the top and curved down to a point below.

Sometimes people make the horizontal cut first. Then they start the shield at the bottom, slicing up until the top straight cut is reached, and being careful to turn the knife left and right at the end to detach the shield all in one smooth piece. If you do cut a sliver of wood—and this often happens because your instinct always will be to avoid cutting into the cambium—you can remove it with ease. With some budding grafts you will have better success if it is removed. With others, no harm seems to be done if it is left on.

But the worst harm will come if you rip out the core of phloem which is needed to supply the bud with food. In your attempt to remove a sliver of wood, never do that. To avoid ripping out the core, make the horizontal cut just down to the wood and no further. Then press the bud shield with the wood attached to it back against the bud stick and try to snap it loose from the wood. If the bark is good and slippery as it will be from mid-spring to mid-fall, you can then slide the shield sideways and leave the wood behind. When you merely pull it off, the core of phloem stays with the wood, and you make an empty, lifeless hole in the bud and ruin it.

After matching the cambiums, or vascular systems, of the prepared scions and the cut rootstocks so that they come together exactly, take good care of the new grafts. Keep the moisture well up, and if there is a hot spell, prepare to shade budded plants to prevent wilting. (For protection you can also make grafts on the side of the plant away from hot sun, preferably to the north or east.) And you should shield the plant from wind. The binding or tape or whatever you use to hold the two pieces together can help, of course, in protection. Though oxygen is needed in the area of the joining, light is not.

Upside-Down T-Budding

In some climates it is better to use an upside-down T in making your graft. Where there is a lot of rain at the season of budding, the water can run right down to the wound in the stock, and can get in there under the tape or rubber band and start a place for rot. There are plants which bleed, too, and the upside-down T provides better drainage than the right-side-up T. If you decide to use the upside-down T, and therefore reverse the cut made in the stock, reverse also the way you make the shield on the scion. See that it is upside-down, too, with the horizontal line at the bottom, not at the top. Put the bud in from the bottom of the cut and push it up, not down as you do in regular T-budding. The aim in both is, of course, to get the places where there are straight cuts together. Naturally you wouldn't think of upsetting the polarity of a plant by putting a right-side-up shield into an upside-down T.

Patch Budding

A second variation on the T-cut is to remove a rectangular piece or patch of bark from the stock plant. Then you cut a patch piece around the bud to match exactly. Both barks should be in a condition of slipping easily. This method is used on rubber trees, and on walnuts, pecans, and other thick-barked trees which do not lend themselves to neat T-cuts. The method is also used sometimes for materials that are quite different in size. The bud may come from a ½-inch stem and be put into a patch cut on a 3- or 4-inch stem. Even so, the general rule for budding is to have both stock and scion of just about the same caliber.

Several cautions are in order if you do patch budding. One is to use the best tools possible: in the earlier section on equipment for grafting, I mentioned a very handy 2-bladed knife that can be used to make the cuts. Another caution is not to make the bark piece for the patch more than 1 inch wide. Also, when you slide the bark out for the bud piece, very carefully slide it out sideways, not up or down, because then you might injure the core of the bud. And above all, be sure that the 2 patch shapes fit exactly. If the bark of the stock is thicker than that of the scion, pare down the stock bark a little. Otherwise, when you put on the wrapping, the bud patch might not be strapped tightly enough and water might

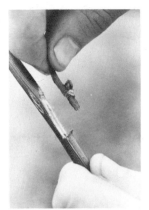

get in. Never delay in putting the 2 patches together. The barks must be moist and fresh. It is possible to cut the slits through the bark of the stock plant a week ahead. This not only saves time, but also starts the formation of callus so that when the stock patch is removed and the bud is added, healing can proceed more quickly.

A final caution: since these young materials are growing very fast, keep track of the tape or rubber band around the union. If you do not cut it in time, it can constrict the stock plant and ruin your undertaking. Merely sever the wrapping on the side opposite the bud. Do not remove it. If you tend to be awkward or careless, put some extra thickness under the tape so that when you slit through the wrap, you won't injure the tender bark underneath.

A third variation on the T-cut is to make 2 horizontal slits in the bark of the stock and to connect them with a straight vertical slit, thus making an I-cut. This gives the advantage of 2 side flaps to raise so the bud is easily inserted. It is a method used when the bark of the stock is somewhat thicker than that of the scion but when it is unwise to pare back the stock bark. With an I-cut, such paring is not necessary, for you can merely insert the bud piece flat against the stock, and tie it in.

In very early spring before the bark slips (or in late fall when it no longer slips) you can chip into the rootstock bark

(1) To make a chip bud you prepare the stock by removing about a 1-inch piece of stem. Do this by making a 45° cut just below a bud about ¼ of the way into the stock. (2) Then make another 45° cut an inch above the first cut (and above the bud). Cut down until you reach the first cut. Lift out the "chip." Do not slice into the cambium. (3) Remove a bud from your scion by cutting it the same length and at the same angle as your stock cuts so that the scion bud and stock will fit together well. (4) Tie or tape the graft and cover well with wax.

I-Budding

Chip Budding

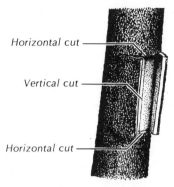

Horizontal cut

Vertical cut

Horizontal cut

Make two horizontal
slits in the bark
of the stock and
connect them with
a straight vertical cut,

the two side
flaps raise
easily for
inserting the
bud piece,

and tie
with tape.

I-Budding

between the nodes and prepare a place to put in the scion
bud. First make a cut at a 45° angle, starting just below the
bud and going down into the wood. It will probably reach ¼
of the way through the stock. Then make a second 45° angle
cut, starting about 1 inch above the first cut. Slice down
through until you reach the first cut at its innermost edge.
Remove this cutout material. Then as well as you can, make
similar cuts into the scion bud stick. Start ¼ inch below the
bud for the first cut at a 45° angle. Begin the second cut ½
inch above the bud, going down behind it until your knife
meets the first cut. Then fit the 2 pieces together and strap
them. This method is favored for budding grapes.

Other Variations   Cutting techniques used in top-grafting and double-
working are discussed in the later section on the main types of
grafting. When you use smaller scions and do top-budding,
your methods and aims are just about the same.

In the method called *microbudding*, even smaller pieces
are used. Only the bud, with a tiny sliver of barkless wood is
slipped into an upside-down T-cut on the stock. The whole
graft is covered with polyvinyl chloride plastic, unless you
object to using such materials. I can see that there may be

some good commercial reasons for attempting microbudding, but for the amateur gardener certainly bigger, more manageable and less finicky materials are to be preferred.

When to Do Bud Grafts

Fall Budding

As I mentioned earlier, if the rootstock you are using is a seedling of a fast-growing species and one season's growth is adequate before budding it, you can do the budding in fall. Use a bud stick from the current season's growth and do the budding when the temperature is above 45°F. and under 90°F. The optimum seems to be about 80°F. When the callus has formed and water and nutrients are moving up and down the rootstock and somewhat into the new bud, bud and stock may unite in 2 or 3 weeks. You can tell sometimes by the condition of the petiole left on the bud piece. If it drops off, the union has taken place; if it merely shrivels and darkens, the union is not taking place.

Do not expect bud growth in the fall, however. Dormancy is what you want during a cold spell, and spring is early enough for progress. Once in a while buds do grow and are winter-killed. This can happen with roses and honeylocusts and occasionally with the swelling buds on maple grafts.

When spring comes to fall-budded grafts, the top of the stock piece can be gradually cut back and off. Make sloping cuts away from the bud, and wax the cut if necessary. If the bud is surely growing, the top of the stock can be removed in 1 cut. With citrus plants, the top can be merely bent over or partly cut through in hopes that the leaves of the stock plant will still provide a little food to the growing new bud. If you live where it is very windy, it may be better to leave the top on to use as a stake for the spring shoot until it is strong enough to hold itself up. Otherwise use a stake to help the new shoot. If latent buds develop on the rootstock, it is best to rub them off and let all the strength of the plant go to the new bud and shoot. When all goes well, at the end of the season the plant with a 1-year-old top and 2-year-old stock will be ready to move to its permanent place.

June Budding

More speed of development can be had by doing the budding early in the season, when the inserted bud can be

For June budding remove bud from the current season's growth to use as the scion (1). Cut a shield in the young rootstock the size of the scion bud, and insert the bud (2). Then tie the bud graft and remove the top leaves of the stock.

forced into growth right away. Though this is not sensible where the summers are short, in Florida and adjacent states or in warm parts of California June budding is very helpful, especially in the propagation of fruit trees. It has been used with considerable success with stone fruits, including almonds, apricots, nectarines, peaches, and plums, using peach seedlings as rootstocks. To do June budding, plant seeds in the fall (or stratified seeds very early in the spring), so that the seedlings are tall enough to use as rootstocks by mid-June, the latest date preferred for this type of budding. They will be quite small, however, often not much more than 6 inches tall and ⅜ inch in diameter.

The bud sticks also should be of the current season's growth, mature enough to have good buds in the axils of the leaves. Since they will be nowhere near their dormant period, you can expect active growth from the moment they are inserted in the stock. To help matters along, before insertion remove any sliver of wood that you may have gotten when you cut the shield on the bud stick.

Because the two parts of the new plant are so young, it is better not to remove many leaves. Let them remain on the stock even below the point of grafting, which should be about 5 or 6 inches above ground.

With June weather and the youthfulness of the plants, growth of the grafted bud may well begin within a few days or a week after insertion. Begin to cut back the top part of the stock seedling 1 week after budding, reducing it by 1 inch or so each time you cut. Ten days after the original budding, cut it back again. Finally cut it a third time, right back to just above the union. At the same time (earlier if necessary) take off any new sprouts that appear on the stock. They sap the strength that you want to go into the grafted new bud, but they arrive, of course, because you have removed the top growing tip which usually would be dominant and able to limit any side shoots.

Spring Budding

Spring budding is not done until the seedling or cutting to be used as stock has had a whole season to grow plus a winter's rest. But do it early before the spring upsurge of growth has really begun. Use the same kind of buds you would for other seasons, seeing that they are full, healthy, and vigorous. Collect the bud sticks while they are still dormant, and store them in the refrigerator if necessary to be sure the growth is arrested.

Scion wood for spring budding can be taken from big stock trees (or small woody stock trees) during the winter and stored until the understocks are ready to receive it. You can do this anytime you are sure that the material is not frozen and doesn't show any freezing injury. Select vigorous, disease-free, well-nourished stems from the center or basal part of a branch, which will be high in stored carbohydrates. Take the stems from shoots which made up to 2 or 3 feet of growth the previous season. Since most grafting is done with plant improvement in mind, the material should also be genetically reliable and sure to breed true.

When you have shoots ¼ to ½ inch in diameter all ready to store, wrap them in barely moist peat moss or slightly damp shavings. Then cover the bundle with polyethylene plastic (the kind which permits passage of gases but not water). Store at around 40°F. in the refrigerator or cold frame for 1 month or less. Look at the shoots once in a while to see whether buds are starting. If so, reduce the temperature to just about 32°F. to assure continued dormancy. If after a month you still need to continue storing the scions, the temperature probably should be reduced anyway to be on the safe side. Never use the freezer; that is too cold. (If in spite of all your efforts buds still do come, throw out the sticks or use them for cuttings: grafts will not take if the scions are leafing out because transpiration from the leaves will use up all the water and nutrients from the scion and make it die.)

Keep the bud sticks dormant until the rootstock plants are really growing. Then as soon as the bark on the rootstock is easily slipped, insert the new buds and tie them in place with a tape or rubber band. The union and growth will take longer than in June, so postpone the cutting back of the rootstock above the union until 2 weeks after the grafting. Take off the rootstock's own buds whenever you see them during this period unless you really need them to shade the union area or to make food in their leaves.

There are advantages to budding in spring, as compared to the other two possible times: the greatest of these is the long growing season ahead. Growth is also rapid and vigorous once it gets started. The disadvantages are that the weather is not so warm and favorable as in the later seasons, and that it's a bother to store the bud sticks until the stocks are ready to receive them.

Grafting

In grafting, the diameter of the stock usually will be greater than in budding, and the scion will be larger than in budding and likely to have several buds instead of one. Needless to say, both parties in the graft should be as vigorous and disease-free and genetically reliable as materials used in budding. Grafting may be done in spring, summer, or fall, though the types which require bark to be slipped are best done in mid-spring. (With some house plant materials you can make grafts in any season.) If you plan to graft in mid-spring, keep your scions dormant until the rootstocks are ready for them. Because of the wide range of grafting materials and situations, there are many, many ways in which stock and scion can be prepared for a graft union. As you'll see, the kinds of cuts made often give names to the various methods.

Whip (or Tongue) Grafting

Whip grafting is a common method, frequently used for small, young ¼- to ½-inch material. It is very often completely successful because with the V-shaped cuts you make, a considerable amount of cambium contact is provided, especially when the diameter of the stock and scion are equal. You should make the V with 2 cuts on the stock. The first is a long, smooth diagonal cut of up to 2½ inches right through the stem, executed from below to above with a very sharp knife. Your second cut from above to below should make a very narrow V. It will be only about half as long as the first cut, and will end at a point about ⅓ of the distance down the stem from the tip of the stock. Then remove the small,

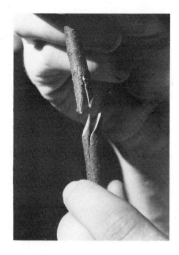

remaining V-shaped wedge of wood. (See illustration.) Prepare the scion with cuts of exactly the same size and angle on a bud stick between 4 and 6 inches long, cutting first down and then up, in reverse cuts so that the stock and scion will fit. For ordinary whip grafting the scion should have at least 3 eyes (or undeveloped buds), though scions with 2 eyes or even 1 have been known to take. When you fit the stock and scion together, do not let the lower end of the scion overlap the top of the stock very much or there may be extra callus growth there which will hamper the final healing. The same thing may happen if the stock is too big for the scion. Since there will rarely be an absolutely perfect fit, hold the graft together with tape or waxed string, and apply wax or wrap in damp peat moss or sawdust to keep the graft from drying out. If, however, the grafted pieces are to be planted soon, you can put the graft area below ground. Then protection from drying will not be needed, for the soil will keep the area moist. Unless underground, the tape should be removed as soon as the graft has taken, for fear the tape will get too tight as the plant grows and strangle it. The adhesive tape used in nurseries is usually put around tongue grafts when they are used all around a tree in the practice of topworking. This, too, has to be cut off when the new growth starts. In all these operations avoid splitting the grain of the wood.

(1) Prepare your stock for a whip or tongue graft by making 3 cuts. The first is a diagonal cut right through the stem you're working with. The second is about half as long as the first and runs lengthwise right through the middle of the first. And the third cut is made not quite parallel to the second, but running into it to form a V. After this third cut you should be able to remove the wedge from the stock that's formed by the V-cuts. (2) Make cuts the same size and at the same angle on a 4- to 6-inch long bud stick to form the scion. (3) Fit the scion into the stock and when you're sure the cambiums are meeting, tape and then wax the union. The lower end of the scion should not overlap the stock very much or extra callus will form there and that will discourage proper healing of the graft.

Splice Grafting

In splice grafting you make just one smooth slice at the top and bottom of the two pieces to be grafted. This method doesn't permit such a tight fit, but it's very useful for herbaceous plants and for plants with very pithy stems or wood so stiff you can't make a nice flexible tongue cut. It is quite easy to do and can be tied as you do a whip graft.

Side Grafting

Of the several kinds of side grafting, first I'll discuss *side-tongue* grafting, used for branches of trees which are too large for a whip graft. You make an oblique cut into the stock at 30° angle and a similar cut on the scion, to fit. This is a bit tricky because the scion in side grafting is likely to be a good deal smaller than the stock. The best fit is often attained by

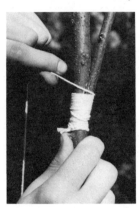

(1) To make a side-tongue graft, make 2 parallel, diagonal cuts into the stock. The first cut should be made so that you remove up to 2½ inches of the bark. The second should begin ½ inch or so above the top of the first cut and end ½ inch above the bottom of the first. It should cut right into the stock. The two cuts should not meet. (2) Prepare the scion, taken from a bud stick, making cuts exactly the same size and at the same angle as those of the stock. (3) Then insert the scion in the stock, angling the scion so there is a tight union, with cambiums matching. (4) Hold the graft together with tape or waxed string and then cover it completely with wax.

insertion of the scion at an angle, especially if you have a 3-inch, 3-budded scion, for instance, to fit 1 inch down into the cut branch. As long as the cambium areas touch, the graft has a good chance of being successful.

A *spliced side* graft or *veneer* graft, so-called, is often used on seedling evergreens. Here you make a smooth downward and inward cut of about 1 to 1½ inches in the stock plant, beginning the slice just above the crown in a smooth area, and ending a few inches below, where you make a short second inward cut. The idea is to make a wedge of wood and bark to take out. Then the scion is given 2 similar cuts, made so that the cambium layers may be matched exactly. When

the graft is fitted, wax it; or wrap it in damp peat moss; or put it under mist or in a high humidity case, depending on conditions. (Evergreens are more reluctant to form callus than deciduous plants, so special measures like misting may be needed with them.) As with other methods, when the graft union finally heals, cut back the stock to let the scion take over.

The cleft graft is used on plants larger than 1 inch in diameter. After you cut off a limb, divide the stub (stock) by chopping or slicing into it to make an opening of just the right size so that 2 small scions may be inserted in the cambium area when the cleft is held open with a wedge. Each scion must be carefully tapered to fit, and you should insert one in each side to keep the cleft balanced. Wax the entire cut surface of both stock and scions, for there is quite a bit of dangerously exposed surface with this kind of graft. When the grafts take, the one that is not the better one can be cut away. In many instances, if you left both on, there would be a weak crotch. When 1 of the 2 scions dies, use a sharp knife to cut it off, at an angle, and apply wax to this new wound. (If both scions die, cut or clip both off and permit a water sprout to grow up as a new site for grafting.)

Prepare your scions for cleft grafting when the wood is dormant, but you'll get best results from cleft grafts made in the spring when sap in the stock is rising.

A variation on the spliced side graft and the side-tongue graft is the side-stub graft shown here. (1) The stock is prepared by making a diagonal cut ⅓ to ½ of the way through it. Then (2) the end of scion is cut into a narrow wedge. (3) The top of the stock cut is pulled open and the scion's wedge is inserted so that the cambiums match. The (4) the stock is cut off just above the union, and the entire graft is waxed.

Cleft Grafting

Notch and Wedge Grafting

When the branches are as big as 2 to 4 inches in diameter, a more shallow split may be made. This so-called notch and wedge graft is very useful for curly-grained woods. Make

Prepare the stock for a notch and wedge graft by cutting a wedge almost to the center (1). The scion should have two buds, 4 to 5 inches long, and should be tapered to fit right into the notch in the stock (2). Then wax the graft well (3).

three cuts with a thin-bladed saw almost to the center of the stock, and then widen the cuts so as to make perfect fits with the prepared scions. The scions should be 4 to 5 inches long, with two buds. Prepare each one by shaping a long tapering end to fit the notches in the stub of the stock. No tying is needed, but wax exposed surfaces as in the previous method. V-cut wedges can also be made and the scions cut to fit.

(1) Start a cleft graft by cutting off the upper parts of a few branches 3 or 4 inches in diameter. These branches, which will become the stocks, should be equally spaced around the tree. (2) Split the stub (stock) several inches, just big enough to hold two small scions when the split is held open with a wedge. (3) Then prepare the scions by cutting one end of each into a gradually tapering wedge. (4) Hold the split in the stock open with a wedge and insert the two scions, one at each end of the split. Make sure that they are positioned so that the cambiums match. (5) When the scions are in place the wedge is removed. (6) The union may then be taped as here, and then all the exposed surfaces waxed.

Bark grafting is much simpler than the methods just described. It causes little damage to the plants, and it also allows quick healing because of the excellent cambium contact that is possible. With it, you cut off a limb of the stock plant, then

Bark Grafting

Bark grafting is a simple method for grafting several scions to the stump of a main limb of a tree. (1) After you have cut off the limb that is to be your

make a slit in the bark from the top of the stub downward, about 2 inches long. Make 2 or 3 more similar slits around the bark. Prepare the scions with smooth, slanting cuts on one side and with a less slanting cut of about ½ inch on the other side of each. Next, pull back the bark on both sides of this slit, being careful not to tear the bark. The scions will then go snugly into the openings in the bark. Fit them in with the long wedged sides against the inside of the stock, and then nail down the flaps of bark on the stock or tie them in position so

stock, make a slit in the bark beginning at the sawed-off edge and continuing straight down the limb about 2 inches. Make two or three other similar slits around the bark. (2) Prepare the scions so that one side of each has a long slanting cut, forming a wedge-like shape, and the other side has a more gradual, shorter cut that forms a slighter wedge. (3) Pull back on the bark on both sides of the slit without tearing the bark and place 1 scion, long wedge against the inside of the stock, into each slit. Nail or tie the flaps of the bark closed over the scions. (4) Then cover each graft and the sawed-off edge of the stock with wax.

the scions and stock are snugly put together. Cover the whole area with wax.

A second method of bark grafting is often used in spring when the bark will slip easily. Here you cut into the bark just enough so that a section can be lifted to slip in a scion. An advantage of this kind of bark grafting is that you need make only 1 cut in the bark. And if the stock branch is over 2 inches in diameter, there will be room to slip in 2 scions on the opposite sides of the branch.

Approach Grafting

In the quite different method of approach grafting, two independent and living plants are pulled together and grafted after you slice off sections of bark where their branches are going to make contact. Usually one or both of the plants so grafted is grown in a container. If the rootstock plant is in the

Approach grafting is used to unite two separate, living plants. One of the most common types of cuts for this graft is the side-tongue cut. (1) and (2) Bring the two plants close to one another (one is usually growing in a pot) and cut 2 parallel and diagonal cuts in each just as you do for a side-tongue graft. (3) Then bring the stock and scion together, fitting one side-tongue cut into the other. (4) Once you're sure the union is tight and the cambiums are touching, secure the graft with waxed string or tape, and then cover with wax. After the graft has taken, cut off the top of the stock plant.

container, move it near to the scion plant, preferably in the spring when growth is active, though this kind of grafting can be done later in the summer, too, even in winter with indoor plants. To attain a very tight fit, you can use the tongue or whip cuts for approach grafting; or you can cut a slot in a thick-barked stock plant just the width of the small scion, then tie the scion into the slot, nail it down, and apply wax. After the graft takes, remove the top of the stock plant for a cutting, and allow the scion to take over and create the new independent plant.

Bridge grafting is used for trees which have been injured or girdled by rodents (as often happens if people neglect to protect young plants in the winter with wire around the trunks). It is also helpful for plants damaged in winter storms. To keep the trees from dying, cut scions from young growth on the tree or from a tree of the same variety, making them 1 or 2 inches longer than the damaged area to be bridged. Take the scions from healthy, 1-year-old-growth with a diameter of from ¼ to ½ inch.

Allow 4 or more scions for each limb to be grafted, or enough so that there will be one to every 2 to 3 inches of the circumference. But allow more if you are bridge-grafting a trunk. Next, cut both ends of each scion to a wedge-shaped

Bridge Grafting

taper, and fit them exactly into 2-inch slits made in the bark on the stock above and below the damaged area. Leave ½-inch flaps of bark which can be nailed down and waxed after the scions are inserted.

For best results you should prepare the damaged tree by trimming the bark back until you reach undamaged tissue. Take out all jagged pieces.

This kind of graft should be done in the spring, fairly late so that—contrary to usual practice—the buds on the scions have started to grow. Remove the buds, and make sure the lengths

Begin the bridge graft by removing jagged edges of the damaged area of the two slits in the bark above the damaged area of the bark on the stock (1). And then make 2 slits in the bark above the damaged area and 2 below to form flaps in the bark. Cut scions an inch or two longer than the damaged bark area and cut both ends of it into wedges (2). Slip one end of the scion under the bark flap above the damaged area and the other in the flap below (3). Repeat this same process all around the stock at least 3 more times so that there are 4 or more scions inserted both above and below the damaged bark area of the stock. Then tie and wax over all cut surfaces.

of scion arch out from the branch or trunk, for this bowed position helps to insure good, firm contact between the stock and the new scions. Secure them well and cover all cut surfaces with wax, including all surfaces of the original injury. This will help to prevent disease and decay in the bridge graft, which is particularly open to trouble. As buds on the scions appear, keep removing them until all areas are healed and good growth has been established. Grafting like this is very valuable for saving trees when cambium and phloem have been destroyed all the way around the trunk. Such girdling means that the plants have no means of sending carbohydrates down from the leaves to the roots, and they will die if they are not grafted.

Inarching

Another form of repair grafting is inarching, used especially for improving the root system of old damaged trees. With this method you plant seedlings or rooted cuttings beside the older trees, then graft them into the trunk to provide the old tree with a new set of roots. Plant these seedlings about 5 or 6 inches apart around the circumference of the tree, during the dormant period. Then, during the first period of active growth in the spring, start grafting. Cut slits just the width of each scion seedling through the bark of the stock tree. Then make a long shallow cut along the scion on the side next to the tree. Next shape the scion into a wedge at the end by cutting off about ½ inch on both sides. Then put this tip in under the bark of the stock and tie and wax the whole area.

Double-Grafting

Although there are many plants which are incompatible, an in-between graft can sometimes produce compatibility if the plant material in the middle is compatible with both stock and scion. Make the first addition to the rootstock by whip grafting or budding. After a season's growth, make the second graft to the scion, which now has become the stock. By resolving incompatibility this way, you can provide a cold-resistant or disease-resistant stock for a desired tree; can get strong trunks or strong, wide crotches; or as is most often desirable, you can get dwarf trees. When quince stock, for instance, is used to dwarf 'Bartlett' pears, it is necessary to introduce an interstock in the form of another pear, for 'Bartlett' is not compatible with quince. The famous 'Malling IX' is used to dwarf apple trees as an interstock grafted onto a lower stock of a different Malling variety. This method takes longer than simpler methods, but is well worthwhile to get the fine results that can be reached. (See fruit entries in Part II.)

Top-grafting is more an approach to grafting than it is a type of grafting. What you do in top-grafting is prune back several branches of a tree and graft new limbs onto these stubs. It is considered most appropriate for long-lived trees that the propagator wants to change after they have attained mature growth. You start with 3 or 4 small, secondary scaffold branches that are 3 or 4 inches in diameter. They should be well spaced around the tree and up and down the trunk. First, cut the branches by the safe 2-cut method: cut down through the branch, about 1 foot from the trunk, sawing about ⅓ of the way through it. Then saw upwards 1 inch or so nearer the trunk until the branch drops off. Then smooth and divide the cut area on each branch and insert 2 scions. As in cleft grafting, you will cut off the weaker one after the graft has taken.

In top-grafting you can use bark, whip, cleft, side, or notch and wedge grafting, preferably in the spring. In fact, if you use bark grafting, you must do it in spring while the bark is still slipping easily. When you do top-grafting, sometimes the severe pruning of the old tree, which finally involves removal of all old branches, creates danger of leaf burn and even death. If work is done when the weather is still cool and the tree still really dormant, severe injury is less likely.

Once in a while after a tree is entirely cut back, the flow of nutrients and water to the scions overstimulates them, and they grow so succulent and soft that they die during the following winter. It is better not to prune so drastically, especially on the south and west sides of trees that may be subject to sunburn. Among those susceptible are citrus trees

Top-Grafting or Topworking

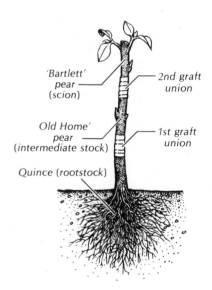

'Bartlett' pear (scion) — 2nd graft union

'Old Home' pear (intermediate stock) — 1st graft union

Quince (rootstock)

The three distinct parts and two-graft union in double-grafting.

and olives. With them you should space your pruning over a period of 2 or 3 years. Another way to avoid trouble is to use younger rather than older trees for topworking or top-grafting.

Double-Topworking

Double-topworking is the same as double-grafting, except that you are resolving the incompatibility with the desired scion up at the top of the tree. This method also requires 2 years' growth while first one graft and then the next are becoming adjusted. Again you are advised to use younger rather than older trees and to avoid grafting all branches in one year. Certain additional precautions should be taken. When your tree is a walnut or other bleeder, make some extra cuts to let the juices out there so they won't all try to come out at graft wounds and clog up the grafted area. Keep watch over grafted trees, and if some of the scions die, remove them at once with clippers or a sharp knife. Wax again if surfaces are reexposed. Dip the saw or other instrument in a 10 percent solution of Clorox® to sterilize it. And dip again and wax again whenever new projects are undertaken.

Cutting Grafts

A very handy kind of grafting will allow you to join a cutting with some leaves left on and a small scion, also with leaves. (But these should not be tender, delicate leaves that

For a cutting graft, take a cutting of an evergreen or citrus tree for the stock. Cut out a splice of the stock on about a 30° angle (1). Prepare the scion with a splice in the opposite direction so that it fits the stock. (2) Tape the graft and put the grafted cutting into a rooting medium and keep the humidity around it high.

may wilt.) This method is suitable for evergreens, either narrow-leaved ones like the difficult conifers or broad-leaved types like the equally difficult rhododendrons. It can also be used for citrus trees. The idea is to make a simple splice cut and insert the scion on a slope of 30°. Then put the new combination into a rooting medium, and treat it as though it were an evergreen cutting. (See Semi-Hardwood Cuttings in chapter 2.) You will probably find it necessary to use misting to keep the plant going. When the healing is complete in about 8 weeks, cut off the top of the stock and plant the rooted cutting with its new top out in the garden or nursery row.

Herbaceous Grafting

Herbaceous grafting is more commonly done by professionals for study purposes than by amateur propagators, though the method is also employed to graft grapes. Propagators use stocks in the seedling stage. Since both stocks and scions are young and soft, professionals rely most often on the simple splice graft. The materials are very young and pliable, so they are subject to flopping and wilting. Instead of a stake, it is the custom to use a small piece of polyethylene tubing slipped over the fitted stock and scion. You make the graft itself merely by giving stock and scion similar diagonal cuts, which are then fitted together. Healing is rapid, and is usually complete about 10 or 12 days after the graft.

Somewhat older herbaceous materials can also be used for grafting, if you vary the method a little. Leave a few leaves on the stock in order to support its life systems while healing is taking place. Make a cleft graft to insure better contact between the stock and scion, using only one of each of the same diameter. Then bind the union area with rubber bands such as those used in budding or use soft raffia strands. Propagators of grapes often use adhesive latex nursery tape, which goes on very easily. And they find that it is best to use a scion with only 1 bud. Stake this kind of graft. Expect healing in 2 weeks.

A simple herbaceous graft has been performed to join these two grapevines. The plastic tubing slipped over the graft prevents the vines from falling over and wilting.

Other Grafts

Professionals at experiment stations even make nurse-seed grafts in seed embryos, using either the radicle or plumule of the tiny structure, soon after the seed has germinated. As with another professional practice of root grafting, the methods are so delicate and special they are not recommended for amateur grafters.

Chapter 6

Seeds and Sexual Reproduction

Since propagation of flowering plants by sexual means is by far the most common method of reproduction in nature, it is important for gardeners to know the basic events and the skills to be learned for this kind of propagation as well as for the asexual kinds. The key to the process is the flower, which produces the fertilized seeds that you plant.

The flower is a specialized reproductive shoot, with male and female parts, all of which contribute to the success of seed production. Flowering comes after the vegetative stage in growing plants or after certain artificial stimuli you can give them. These include girdling, root pruning, pinching off shoots, grafting onto dwarfing rootstocks, or even tying the branches into knots.

The stem of the flower supports a platform from which grow the whorls of the flower's parts. The first whorl is the *calyx,* made up of *sepals,* which protect the bud when the flower has not yet opened. Though usually green, sometimes they are colored, as in tulips. The next whorl is the *corolla,* made up of *petals.* Since these two whorls are not involved in the essential reproductive process of fertilization, they are called *accessory parts.* They are often removed during the process of hand-pollination.

The Flower

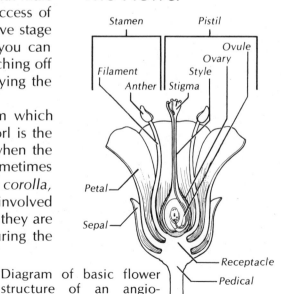

Diagram of basic flower structure of an angiospermous plant.

117

A cross section of a zinnia.

The male parts of the flower, the *stamens,* are the third whorl. Each stamen has an *anther,* which produces the pollen. In some flowers, pollen grains are produced by the millions. They supply the sperm during fertilization.

The stamens usually surround the central, female part of the flower, called the *pistil.* A typical pistil has three parts. The top portion is the *stigma,* usually a flat, sticky surface which is perfect for receiving pollen. The middle portion—long and slender—is the *style,* down which the pollen tube travels during fertilization. The swollen basal portion is the *ovary,* made up of one or more sections containing one or more *ovules,* each of which contains an egg. After pollination and fertilization the ovary becomes the *fruit* and the ovule becomes a *seed.* Good fruits and good seeds are the usual goals in sexual plant propagation.

Variations

There are many variations in the appearance, size, and structure of flower parts. The petals and sepals of insect-pollinated plants are usually bright, showy, and fragrant, though on the wind-pollinated flowers of wheat and other grains and many trees they are so minute as to be unnoticed by man or insect. Sometimes they are absent altogether.

In some species, the male and female sexual organs are located in separate flowers or on different plants. Squash and corn are examples of *monoecious* plants, which have staminate and pistillate flowers on the same plant. (In corn, of course, the tassel is made up of many male flowers and the ear and silk of many female flowers.)

Dioecious plants, on the other hand, have staminate and pistillate flowers on separate plants, which we speak of as male and female plants. Asparagus, willow, and holly are examples of this type. Obviously, only the female plants produce fruits and seeds after pollination and fertilization.

Other flower variations involve the structure of the various whorls. Sometimes, as with the sunflower, the anthers are united around the pistil in the form of a tube. They shed their pollen down into the tube, so as the pistil grows up, it has to push through it and expose the stigma. Then a pollinating insect brings pollen from another flower and thus effects cross-pollination, which is usually beneficial to the species. Sometimes, as with grass of Parnassus, you can see little structures which look like a second set of stamens. They have no pollen, however, and are just there as honey-guides to help the bees. With morning glories, bluets, petunias, foxgloves, and blueberries, the petals are partially or completely united into a *corolla tube*.

Nectaries are especially important modifications, for these glands produce the nectar that attracts bees, wasps, and other insects. Nectaries are found on the receptacle, along the petals, or deep within the spur of the flower. Some "petals" are really colorful specialized leaves called *bracts*, which appear to function as petals. In the flowering dogwood and poinsettia, these bracts surround a cluster of tiny flowers.

Causes of Flowering

What brings on flowering is still something of a mystery, but plant physiologists have learned a great deal about hormones, temperature, and seasonal needs in recent years. Perhaps the most well-known discovery is that day length affects the flowering of certain plants. As I suggested in chapter 1, actually it is the length of the dark period that is important, for the leaves seem to sense this. When the day length is right for them, they produce the hormone (or combination of hormones) which induces the apical meristems to produce flower buds. Examples of short-day plants—or long-night plants, as they really should be called—are chrysanthemums, cosmos, Christmas cacti, poinsettias, soybeans, and *Kalanchoe blossfeldianas*. Long-day plants include dill, plantain, beets, and various rudbeckias such as the black-eyed Susan and coneflowers. Day-neutral plants we are familiar with are tomatoes, beans, and corn.

To induce short-day plants to flower, you must keep them in the dark each day for the length of time approximating a night in October or November. Even a flash of light in the middle of the dark period is enough to prevent the formation of flower buds.

Flowering can also be induced by a period of low temperature, a process called *vernalization*. Cabbage, celery, and beets are examples of plants which require vernalization. A few plants such as spinach, rice, and China asters require a period of high temperature to induce flowering.

Nectar

The nectar which attracts the bees and other insect pollinators is a sweet and sometimes gummy substance made up of sugar, volatile oil, water, and other components. Once in a while, as with some jasmines and heaths, nectar is poisonous even to bees. The secretion is released by the flower during hours when the temperature rises to between 80° and 100°F., usually in the afternoon, when bee activity is also at its height. Nectar flows best on warm days after cool nights, and the amount increases when the air is somewhat damp. Dull, rainy days decrease the flow. If you're watching for favorable pollinating periods for plant breeding, therefore, be especially alert during good weather.

There are two kinds of nectar, and so-called nuptual nectar is more plentiful during the sexual maturity of the flower's pollen and ovules, which is, of course, the right time for the bees to come. The flowers on cotton, golden currants, horse-chestnuts, and other plants also change color then. The bees notice this and come, for though they are color-blind to red and purple, they see blue, yellow, blue green, and ultraviolet.

The flow of nectar results from a modification of the cell walls in the nectaries in the lower part of the flower. This change brings about a bursting of the cellulose, which then disintegrates into sugars. These sugars are dissolved in water, and osmotic pressure (similar to that which keeps a plant crisp and turgid) easily allows this mild solution to pass out into the tube of the flower.

Rain can stop this flow of nectar, and some flowers close right up or change their position to protect themselves from the rain. Again by osmosis the additional water will cause nectar to spill out of the flower through ruptured cells. When the flower dries again, that loss of nectar stops. On the next good day you can expect to see the bees back again, out on the hunt for the renewed supply.

Pollination

Pollination is the transfer of pollen from a stamen to a stigma; it is not to be confused with fertilization, which occurs later and is the actual union of sperm and egg. The transfer can be carried out by insects, animals, wind, a paintbrush, or your own fingertip.

When you watch bees visiting a flower, you see that they are after both nectar and pollen, and sometimes you can observe them storing pollen in pockets in their hind legs to carry it back to the hive. The signal goes out to the bees to concentrate on one species or so per day. And so they neatly gather ripe pollen from one flower and deposit some of it on the stigma of another flower of that same species nearby.

When a bee visits a flower to gather nectar for her hive, pollen from the flower's stamen sticks to her body. As she makes her rounds some of this pollen falls on the stigmas of other flowers, and *voilà*, pollination has been performed!

Butterflies, moths, beetles, hummingbirds, and other birds do the same, and flies can pollinate too. Primroses are pollinated by the bee fly, and one unusual flower in India, which is warmer than the air and smells rotten, is attractive to a buzz fly which likes such smells. In Africa, even bats are useful in pollinating the baobab and sausage trees.

One amazing kind of pollination by an insect takes place for the cattleya orchid. This flower has pollen pellets with fuzzy tails, called *pollinia*. When the bee comes to the platform, she may get two or more pollinia on her back because she already has touched a very sticky disk. The pollinia load down the bee, which then flies as best she can to the next flower. She there meets up with the very shiny, very sticky stigma, so as the bee goes down for nectar, the sticky fluid clings to the pollinia and they are torn off to pollinate the flower. Leaving, the insect may carry away pollinia from this

The insect-inviting beauty of the cattleya orchid.

121

flower, then the next, and then the next, aiding pollination over and over.

Another more adventurous orchid actually shoot out its pollen at passing bees. This orchid makes doubly sure of pollination, for it has a platform as well for any bee that comes in. A few flowers trap bees or other insects in a sweet, alluring liquid, and make the bees pollinate them during their struggles to get out.

Wind-pollinated flowers need not, of course, be so sweet or so colorful and attractive. Their pollination depends on the haphazard arrival of wind-blown pollen grains at the stigma of the female part of the flower. But nature compensates for possible hazards. The number of pollen grains put forth by one flower may be up in the thousands and in one season a single sorrel plant may produce 393 million pollen grains. Their coats are very hard, sometimes so hard that botanists have not been able to crack them in order to study what's inside. Pollen grains can survive for centuries in the acid, sterile environment of a sphagnum moss bog or in the dryness of deserts. In the Sahara, for instance, they indicate that ages ago pine and oak trees were growing in that very area.

Pollen grains in nature are not viable for very long, sometimes the fertilization span is only a few hours. But orchardists and plant breeders can keep pollen viable at temperatures slightly above freezing for a matter of weeks, for a year in some instances, and even for up to 4 years in the case of the blueberry.

Self-Pollination   When the pollen grains simply move from the anthers of a flower to its own stigma, *self-pollination* occurs; but the term is also used when pollen moves from one flower to another on the same plant or to a flower on another plant developed asexually from the first plant (a clone). In self-pollination, the offspring are very like the parent and their seeds are said to "come true." Examples are barley, oats, wheat, peanuts, soybeans, flax, garden beans, some grasses, peppers, tomatoes, cotton, and a few garden flowers, all of which encourage you to save their seeds.

Cross-pollination is the transfer of pollen from the anther of a flower on one plant to the stigma of a flower on another plant not generated asexually from the first plant. Cross-pollinated plants do not necessarily develop offspring like the female parent, for new genetic material is introduced when pollen comes from a different plant or a different clone, and the resulting seeds do not breed true. Examples are very numerous and include alfalfa, clovers, onions, watermelon, hybrid field corn, most root vegetables, cabbage plants, shrubs, and fruit, nut, shade, and forest trees, and most ornamental flowering plants. Cross-pollination also can take place within one family of plants, as when bees visit squash and gourds or squash and pumpkins on the same day. For some plants, cross-pollination is a necessity if they are to set seeds and bear fruit. This, therefore, is the method of propagation most frequently practiced by orchardists and plant breeders.

Long ago animal and plant breeders learned about "hybrid vigor." This is the exceptional strength, disease resistance, and healthy growth often characteristic of offspring of parents not closely related. Hybrid sweet corn is a good example. Usually such hybrids are not used for seed propagation, however, because whether self- or cross-pollinated, they do not breed true and the offspring do not possess the vigor that made the parents so exceptional. It takes great skill as a plant breeder to maintain the desired strain through very careful selection.

There are a number of plants for which self- or cross-pollination is impossible. *Cross-incompatible* plants are those between which cross-pollination cannot occur. *Self-incompatible,* or self-sterile, plants are those in which self-pollination is impossible. In some cases, self-incompatibility may be due to the varying maturation times of the stigma

123

and pollen; when the pollen of a plant is ready, the stigmas are not. In other cases, pollination occurs, but fertilization doesn't occur, and fruits and seeds are not produced.

Hand-Pollination   Many propagators of fruit trees attend to cross-pollination themselves. The methods they follow are well illustrated by the example of propagating peaches.

How to Pollinate Peach Blossoms   The flowers of peaches come on 1-year-old shoots, about 6 to a node. They have a single pistil and about 30 stamens. When the bud is swelling, the pollen and ovules develop; at the time of bloom the stigma is receptive. To pollinate, you collect pollen from buds well-advanced in the balloon stage (that is, when quite swollen, and with the tips of the anthers showing when you press the petals open). One cup of these buds is enough to do 1,000 pollinations. You then rub the flowers through a sieve or roll them between your fingers to remove the pollen. This should be done while the flowers are still turgid. Use clean, sterilized equipment.

The next step is to dry the pollen in glass vials with stoppers of nonabsorbent cotton. These can be held for a year in a deep freeze. To test the pollen after storage, put a few pollen grains on a drop or so of sugar solution; if the pollen is viable, pollen tubes will begin to grow out from the grains in 3 or 4 hours. Do not use beet sugar.

Emasculation of the Flower   Select the flowers you are going to pollinate and remove the male stamens. Cut through the petals just below the point of attachment, or use your fingernails to pinch them off, using a slight twist to remove the petals and stamens. WARNING: Be sure that no pollen is shed inside the flower while you are working. It helps to stay on the windward side of the tree. Luckily no bees ever come to emasculated flowers, but to control stray pollination you can enclose up to 100 hand-pollinated blossoms in a polyethylene bag. Some people enclose the whole tree in a cheesecloth tent on a frame. If you are working with a female, pollen-sterile tree like 'J.H. Hale', you can omit the emasculation and merely place a bouquet of pollen-bearing peach blossoms in a vase and tie it inside the tent. Then introduce a hive of bees and let them go to work.

You can do the pollinating yourself with the tip of your finger, a camel's-hair brush, or the rubber eraser on a pencil. It is quite easy if you do it when the stigma of the peach blossom has fluid on it, showing that it is receptive. If you find it receptive right after emasculation, do it then, though pistils usually have receptive stigmas for 4 to 7 days unless winds have dried them off. The best temperature is the best temperature for bee flight, that is above 60°F.

Variations of Pollinating Techniques With pears, for instance, you collect pollen by cutting and forcing branches of pear in the dormant or early tight-bud stage, 2 or 3 weeks before the tree will bloom outdoors. Watch and harvest the pollen just before the time for it to be shed. When emasculating, use basal blossoms, not the king blossom or the small terminal ones.

Sometimes it is advisable to apply pollen to emasculated flowers all through the blooming season, as is done with plums. Some plums, however, are self-unfruitful and then it is not necessary to emasculate them.

Apricot pollen and some others are so sensitive and so short-lived that it helps to collect them right into a container in a traveling ice box. They are also very cold-sensitive, however, and have been known to perish from contact with cold fingers. (The best time to collect apricot pollen is fairly early in the morning, and since apricots bloom early in the season, you may very well have cold fingers.) Cover emasculated flowers if frost threatens. Avoid rainy days when pollinating.

Because citrus plants flower in two ways, you should remove the undeveloped buds when you cross-pollinate to prevent any self-pollination within the flower; this often happens because the anthers and stigma are so near to each other. The large, early flowers can be picked, and you can use them for pollinating instead of using pollen on your finger. Remove pistil and petals. Storage of pollen in a freezer is not necessary, but keep it dry and cool.

Because of the odd way that the pollen grains fall out of the blueberry flower, pollinating techniques for these plants involve picking several branches of blossoming 2-year-old wood before the stigmas are ripe to force in water. Use mist, and cut the stems under warm water every few days. When flowers mature, rub them between your fingers every day to get out the pollen. Use it fresh if possible, or store in a capsule.

Hand-pollinating a tomato flower, using a small brush.

Sticky or fragile pollen that has to be stored must be kept very dry. The technique for doing this is to put some calcium chloride in the place of storage so it will take up any excess moisture.

Pollination of some nut trees which are wind-pollinated—such as pecans—involves enclosing the flower in a sausage case for a while before receptivity of the stigma, plugging this container with a wad of cotton, and using a hypodermic needle to introduce the pollen, blowing it in by squeezing a pollen-filled bulb.

There is also the very simple method of bringing in some blossoming branches and placing them over white paper on which the pollen may fall. This is used for such wind-pollinated plants as filberts whose staminate and pistillate flowers develop at widely separate times.

Fertilization

Once a pollen grain lands on a receptive stigma some amazing changes occur. As it is germinating, the pollen grain begins to grow through the tissues of the pistil. The *tube cell,* or *tube nucleus,* leads the way, and another germ cell follows. The destination of this remarkable pollen tube is the *embryo sac* in the ovule. Once the embryo sac is reached, the tube cell disintegrates. The other cell has by this time divided to form two sperms. One sperm fertilizes the egg which develops into the embryo plant, and eventually, after the seed itself has germinated, into a new plant. The other sperm unites with a cell which then proliferates to form the *endosperm,* the food supply for the embryo. This kind of double fertilization is unique to flowering plants. Once it occurs, the petals, sepals, stamens, stigma, and style usually wither away, and the ovary—or fruit as it now should be called—begins to grow very rapidly.

At first, the embryo is just a mass of undifferentiated cells, but then the organs grow and take shape; these are a tiny shoot, called the *epicotyl,* and a tiny root, called the *radicle.* One or two seed leaves, or *cotyledons,* also are formed: plants like corn and lilies have one cotyledon, and are called *monocots;* those with two are *dicots.* The endosperm also

Growth of the Embryo

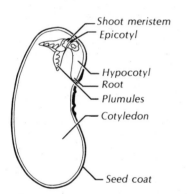

Dicot seed; snap bean showing structures commonly present on legumes.

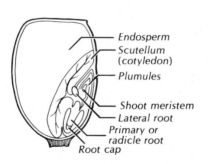

Monocot seed; maize kernel showing structures commonly present in grass seeds.

Four steps in the germination of a bean seed: (1) water uptake begins; (2) seed coat bursts; (3) radicle emerges between cotyledons; (4) plumule appears after radicle has begun its rapid growth.

grows, and in many dicot plants, is completely absorbed by the cotyledons. Peas, beans, pumpkin seeds, and many others have plump, food-filled cotyledons. In all monocots the endosperm is not absorbed, and it makes up much of the bulk of the seed. The edible part of the coconut, for example, is all endosperm. Wheat and corn also have great supplies of unabsorbed endosperm, from which we make flour and meal.

Fruits

Botanically speaking, a fruit is a ripened ovary. The types of fruits are almost as varied as the types of flowers, but there are a few general distinctions. There are dry fruits (pea pods, acorns, walnuts, and grains) and fleshy fruits (tomatoes, peaches, and blackberries). There are simple fruits, which are derived from a single ovary (cherries and beans); aggregate fruits, which develop from several ovaries in the same flower (magnolia and raspberry); and multiple fruits, which grow from the ovaries of several flowers (pineapple and fig).

In some fruits, the fruit coat, or *pericarp,* and seed coat are very close together. Sunflower "seeds" are really fruits, but the inner real seed can be readily separated from the fruit by blue jays, cardinals, or humans. In grains, which are the fruit type characteristic of the grass family, the fruit coat and seed coat are fused.

The period during which fruits and seeds grow is a stressful time for the mother plant. And gardeners are well aware of the need to remove fruits and seed pods as they begin to develop in order to keep the rest of the plant strong and active.

After the seed has reached a certain stage of development, it stops growing and begins to lose water. A mature seed contains very little water, usually around 10 percent, and it is dry and hard. Respiration has slowed down so much that it is not measurable. A seed literally is a little plant in a state of suspended animation.

Clones and Cultivars

When you think of all the hazards and influences surrounding reproduction from seeds, it is not surprising that most efforts to propagate good hybrids depend on vegetative, asexual methods. Familiar clones so preserved include peaches, apples, and pears; ornamentals like dahlias, chrysanthemums, roses, tulips, camellias, and many other herbaceous and woody plants. *Cultivars* are any humanly produced varieties, including those coming from seeds. Cultivars are kept pure by putting flowers in bags or tents. And stock seed breeders keep different varieties a mile or two apart to prevent insect or wind-pollination and to qualify their seeds for certification.

For propagating conifers it is helpful to know how they differ from flowering plants. The majority of conifers have needle-like leaves, are evergreen, and have cones but there are exceptions. A few common conifers like the yew and juniper have a fleshy structure covering their seeds. This is not a true fruit, of course, since only flowering plants possess ovaries, which become fruits.

Conifers are Different

(1) Here are two different cones with their respective seeds. The scales of the female cones spread out at pollination time to catch the pollen grains that float down from male cones above, and after pollination seeds develop at the inner tip of these scales. In (2) you can see more clearly the scales on the cone and the seeds that develop at the scale tips.

In pines, which have a typical conifer life cycle, both male and female cones are produced. The male cones are tiny and short-lived but very prolific; they die and drop off as soon as their clouds of pollen have been shed. The scales of the young female cones spread out at pollination time to catch the pollen grains that float down. A fluid flows from the female structure, touches the pollen grains and goes back again with the grains stuck in it. Then the hole from which the fluid came closes again. In a white pine tree the pollen tube grows for a year as the egg develops, and then fertilization takes place. It is another year before the scales of the female cone open again and the seeds are shed.

Other differences are that conifer seeds have many cotyledons, rather than one or two; and they have storage tissues for food that do not originate in endosperm as in flowering plants.

Spore-Bearing Plants

To assure good germination of firm seeds and spores, sprinkle the seeds on a sterilized brick covered with moist, shredded sphagnum moss. Fern seeds and spores need much moisture so place the brick in a water-filled tray.

Of all the plants that reproduce from spores instead of seeds, ferns probably hold the most interest for gardeners and propagators. Because spores seem so strange, many people hesitate to propagate ferns by this natural means. Nevertheless, growing plants from spores is not difficult, and I'd like to digress from seeds for a few moments to explain reproduction from spores in terms of the propagator's job; you'll find it is a very good way to provide yourself with lots of new little ferns.

The standard old method involves doing some special things to grow spores so they will end up growing more or less the way seeds do. First prepare a brick as explained in chapter 1.

The spores you will sprinkle on this prepared brick are from groups that ripened on the underside of fern fronds in little spore cases called *sporangia*. Watch these through a hand lens to find the moment when the cases open up to shed the spores. Then pick the whole frond and put it on a piece of white paper. When the spores shake out, they may be mixed with chaff, so tap the paper to get the larger pieces to separate from the tiny spores. Clean off the chaff and leave the spores on the paper, then tap them gently onto the moist peat moss or sphagnum moss when the brick has taken up enough water to make the top layer moist.

As they absorb moisture themselves, the planted spores germinate and form a *prothallus,* which is both male and female. If this organ is kept clean and free from disease—and moist—fertilization can take place and new plants can grow. The sperm need the moisture to swim to the egg, so keep drops of moisture on each prothallus at this time.

Sori and indusia of ferns

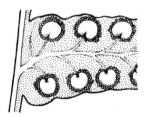

Marginal shield fern

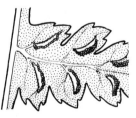

Lady fern

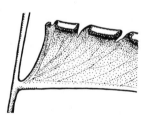

Maidenhair fern

130

Fern spores need light, so merely shake them onto the sphagnum moss. Do not cover. Ten minutes of 100-foot-candle light is sometimes all that is needed. You can use Gro-lux® bulbs or Vitalites®, or put the planted brick on a windowsill. You can use a plastic box for it to keep the moss moist. The time it takes for germination varies from a few days to a few weeks. In 2 or 3 months after you planted the spores on the growing medium, you will see the tiny new fern plants come. You can see them with your own eyes, but if you want the pleasure and excitement of looking them over well, use a hand lens. If the ferns look very crowded, separate them, you can pull some out very gently with tweezers.

During all this time you must watch out for damping-off. Air the plants well if you even suspect that fungus is coming, and put a drop or so of weak Clorox® solution on the infected place. Follow that with a few drops of clear water. Only do a few places at a time; and examine the top of the fern-brick carefully to see that no harm has been done before adding any more. More air and less heat are the safest precautions. As the plants grow, more light is advisable. Two bulbs of 40 watts about 18 inches above the tray will probably be sufficient. Use a Vitalite® or Gro-Lux®. (Ferns are also propagated by rhizome division, as described in chapter 4. Also see Ferns in Part II.)

Although I like the moist brick for growing plants from spores or from tiny seeds (begonias, gloxinias, and African violets), you may prefer the new setup now available from seed companies. I describe it in chapter 1. The advantage of the brick is that moisture is slowly moving up from the tray of water through the brick at all times.

Still another device you can use is a flat container holding water, in which you place a porous clay pot or bulb pan filled with sphagnum moss. The pot can be covered with glass or polyethylene, and sprayed occasionally if necessary.

Ferns Not to Grow   Some ferns that are advertised in wild flower catalogs I would not recommend. These ferns are extremely fussy about their habitat, and most growers are doomed to failure. Even if you grow your own from spores,

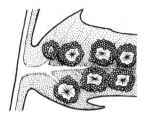

Christmas fern

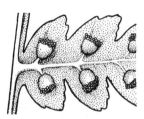

Bulblet fern

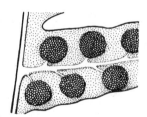

Common polypody

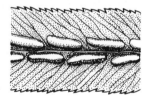

Chain fern

put them out in the prescribed place and tend them as best you can, they may still fail. One I am thinking of is the ebony spleenwort (*Asplenium platyneuron*) which grows on limey, shady, moist rocks on hills and mountains. Closely related is the blackstem spleenwort (*A. resiliens*) growing in a similar habitat, and usually at the base of rocks in the semishade. More rare and difficult is the maidenhair spleenwort (*A. trichomanes*), which needs rocks, a limey soil, and crevices in the rocks to grow in. Mountain spleenwort (*A. montanum*) is particular in another way, for it needs shale, or crevices of sandstone, with only tiny pockets of acid soil. Sometimes people try to grow these in terrariums, but though they may survive for a while, they are short-lived. If you do succumb to the temptation to follow suit, the least you can do is propagate some of these rarities plentifully and persuade all dealers in plant material for terrariums to do the same.

There are quite a few other rarities which you can spot yourself by glancing through a fern guide.

(See House Plants in Part II, for ferns to grow indoors.)

The Wonder of Seeds

Fascinating though ferns and other spore-producing plants are, it is the seed bearers that have contributed most to civilization. "Seeds are the germ of life. A beginning and an end," wrote Orville L. Freeman, former Secretary of Agricul-

The astounding variety of seeds protected by cones, burs, thorns, nutshells, pods, and heavy seed coats and given mobility by wind, water, attached wings, shaking heads, and twisting pods which toss out the seeds when they curl.

ture, "the fruit of yesterday's harvest and the promise of tomorrow's." Students of seeds in labs and experiment stations all over the country, all over the world, continue to remind us that seeds are many things: "a way of survival, a way by which embryonic life can be almost suspended and then revived even years after the parents are dead and gone. Seeds are highly organized fortresses, well stocked with supplies, and are vehicles for the spread of life from place to place, symbol of beginnings and a source of wonder," as Freeman also said.

Everything about seeds—their numbers and forms and structures—relates to their main purpose: to insure continuing life. An orchid or tobacco plant may yield a million seeds in a season. Dry muskmelon seeds can last 30 years; seeds of the evening primrose or mullein can be dormant for 70 years; mustard seeds can stay viable in the soil for nearly a century if the ground is not plowed up; and in an herbarium, seeds of cassia and mimosa have been known to last 200 years. They are so resourceful that propagators hardly need to do more than just cover those they want to grow and keep them watered and aired.

Seeds have a wonderful ability to match their rhythm of growth to the rhythm of the seasons; delayed-action devices, such as inhibitors, impermeable seed coats, and special germination requirements, hold them back as needed. The variety of these devices is amazing. Some seeds have coats that are water-resistant for long periods. To let in some water, the propagator can nick or scratch these seeds as they are nicked by fragments of rock in water passing over them or by the movements of the earth as it freezes and thaws during the winter. Some plants, like the ginkgo, drop naked seeds on the ground where the undeveloped embryos within must grow for many months before the seed is ready to break out of the seed coat. Lettuce seeds, on the other hand, will germinate as soon as they are ripe; but if given a warm moist spell instead of congenial conditions, they will become dormant, as we know when we buy a packet.

Buying Seeds

Seed produced by seed companies or by private growers contracted to grow certain cultivars for them is known as *certified seed*. It is grown in large quantities and approved by a state or federal certifying agency as having a "satisfactory" level of purity, or genetic indentity. *Registered seed* is judged even more strictly. The progeny of the original seed, called breeder's seed or *foundation seed,* is of the highest quality and meets the highest standards. *Breeder's seed* is the parent, originating with the sponsoring plant breeder himself, or with a university or government station which bred the parent plants. These seeds are sent out to seed companies for the production of foundation and certified seeds.

Unless you grow and clean your own seeds, it is wise to get certified seeds so you know you are getting good quality, with a high percentage of plump, viable seeds and little chaff and few weed seeds. You also will get the seeds of the cultivar you ask for, for seeds in this country are labeled in accordance with the Plant Variety Protection Act of 1970.

Buy tree seeds from areas compatible with your own climate and the ecological conditions of your area, including latitude, longitude, and altitude. A few adaptable exceptions include the superior races of East Baltic Scotch pine, Douglas fir from the Palmer area of Oregon, and Norway spruce from the Hartz Mountain region. In buying tree seeds you can also ask for information as to year of production and place of origin (state, county, elevation), and for proof of origin. Make sure that the length of the growing season and the average temperatures during different parts of the growing season in the area of origin are about the same as those of your own locale. It pays to know about the frequency of droughts, too.

(If you collect tree seeds yourself, try to get them from a pure stand so cross-pollination has come from trees of the same type. Detailed information on various species is given in Part II.)

Untreated Seeds

As most people know by now, seed companies add fungicides such as Thiram® to many of their seeds—though there is, as yet, no federal regulation requiring this. You can, however, ask for and get untreated seeds from companies like W. Atlee Burpee, Ferry-Morse, Joseph Harris, Stokes, George W. Park, and Gill Brothers of Portland, Oregon, as well as from such herb specialists as Nichols Garden Nursery in Albany, Oregon; Greene Herb Gardens in Greene, Rhode Island; and Vita Green Farms. And you can save your own seeds.

To sterilize seeds yourself, use a 1 percent solution of Clorox®. Clorox® is already a 5.35 percent solution of sodium hypochlorite, so you dilute the Clorox® to a 1 to 5 proportion of Clorox® and water to get the 1 percent solution (or a 1 to 10 proportion for a .5 percent solution for more delicate seeds such as petunia or nicotiana). Try out a few seeds first to test whether or not they are able to stand as strong a solution as you made, soaking them for 5 to 15 minutes depending on the size and sensitivity. If they look okay, soak the other seeds for that long a time, too. Then wash under running water for 5 to 10 minutes in a sieve or wash and drain at least 3 times.

To be thorough, sterilize your tools and containers, put your tweezers into a flame, and use boiling water or a pressure cooker to sterilize your containers, especially those you use for seeds which you plan to start indoors and which require a long time for germination. For safety's sake you can also work on a counter covered with towels which have been wrung out in a 10 percent Clorox® solution. Also, use boiled or distilled water when you mist the seeds to water them into the sphagnum moss or whatever medium you use. And with all those other precautions, you might as well wear Clorox®-washed gloves, too (especially if you smoke). Some people use a solution of calcium hypochlorite instead of the sodium hypochlorite in Clorox®. But you have to get that in the drugstore and mix it up yourself.

There are many good reasons to save your own seeds; whether you want to breed plants yourself; to increase your supply of plants you like; or to save some time and money by harvesting seeds of garden plants to use the following year. But if you are going to be successful, you must realize that plants intended for seed harvest need to grow under special conditions.

Begin a routine that you can carry over several years, so that after continuous selected harvesting you finally do develop a strain that is well adapted to your own soil and climate. (Professional plant breeders develop strains for a variety of climates.)

Be sure the plants you choose have a record of producing good, big, vigorous, and healthy specimens, outstanding for their species. Plant only one variety of each species to avoid cross-pollination. Fertilize the plants well, and keep them growing far enough apart so that they have plenty of room to spread out and get plenty of sun, and room to send their roots out without interference. Sow early and keep them watered well so that they never wilt and have no setbacks. Pick out the best, earliest, sturdiest plants in the row, and mark them with colored tape or something conspicuous. Do not pick any flowers or vegetables from those plants, but when you see them go to seed, be sure to harvest the first fruits, which should be the most vigorous. The second fruits are often inferior. It is, of course, the quality of the whole plant which you select that will determine the quality of the seed you save. And never work on a wet day.

You may need to put on little bags to protect the seeds of some plants until they are fully ripe. This may also be necessary to save seeds from dropping or popping out as they will from oriental poppies and lupines, for example, or from witch hazel or the fascinating squirting cucumber. Some fruits may just untwist or pop open before you expect them to. This happens with oxalis, pansies, and wild geraniums. Seeds of plants like berries may disappear if birds get to them before you do, but barring this, the fruits are quite easy to harvest and save for seed when ripe. Fleshy fruits, such as those of holly, viburnum, cotoneaster, and kousa dogwood, can be put in water to soak for 2 or 3 days and allowed to ferment. Then you can scoop off the fruit pulp and see the good, plump seeds go to the bottom. Drain and take out the good seeds. Clean again if bits of pulp still cling to the seeds.

As I mentioned earlier, some fruits do not "shed" their seed or seeds. A corn grain or acorn is technically a fruit, not a seed. When dealing with dry fruits which break open and release their seeds, you must be careful to pick the fruit *before* the seeds are discharged. Then shell and clean at once, picking the seeds over and discarding all that are undersized, diseased, damaged, or discolored. Make sure squashes and cucumbers are fully ripe: they will have a hollow sound. If bad weather approaches, or if you're still in doubt about ripeness, you can store these vegetables in a warm, dry room until they are fully ripe. Leave beans on the vine and pull and dry the whole vine before taking out the seeds.

When fruits hold their seeds pretty well, I find it convenient to hang the plants with fruits to be dried in a dry, airy place. Sometimes I stretch them out in the sun for a few days first. I have put such branches in big paper bags before hanging them up. In other instances, I shake out the seeds as soon as they are dry and prepare them for storage. But remember that the least bit of moisture on pea or bean seeds, for instance, encourages fungus attack, so avoid picking such pods when they're wet or early in the morning when they still have dew on them.

Some Seeds Will Not Breed True As I said in describing hybrid vigor, seeds of the first hybrid generation probably will not come true to the parent type; but sometimes—as with geraniums, for instance—who cares? The unexpected pinks, whites, and reds that come are all attractive. With sweet corn it's different: you probably do care. But remember that many of the hybrid corn seeds you buy to plant will produce nonviable seeds anyhow. So save seeds from one of the old-fashioned strains instead. If your aim is to save seeds of a pure strain of corn, plant only that kind for the year you begin to save, because corn cross-breeds very easily. Peas and beans rarely do, but cucumbers of different varieties will, and so will squashes. It's annoying when summer squash and pumpkin cross, as we have had happen. I really don't mind the hard-shelled, rather coarse, resulting vegetable, but I notice that my husband pushes it aside and leaves it on his plate (so I've learned to hide it in a zucchini casserole). Of course the seeds of any such oddities are not worth saving.

Other Vegetables Of the root crops, carrots are the only ones I'd bother to save. Since carrots (and also parsley) are biennials many gardeners would not wish to collect seed from either. But I like to. We leave some carrots in the ground all winter, along with the parsnips, and put some parsley in pots to winter indoors. If I ever found a parsley plant that would winter over outdoors in our climate, I'd certainly save seed from it and try to develop a hardy strain.

Tomato seeds are easy to save. In fact, tomatoes in our garden self-sow so busily that we almost think of them as intruding weeds. One method of using your own tomato seeds is to select a good plump specimen of your favorite kind and just squash it down on the earth in a convenient corner of the garden and wait until the next year. If you don't trust that method, you can ferment the pulp in water for a few days in order to loosen the seeds and then dry and store them until the following spring.

Selecting the Good Seeds When you have harvested the seeds, cleaned them, and prepared them for storage, again select only the best for saving. The good seeds are the plumpest and heaviest, for they have in them the best sources of carbohydrates, fats, and proteins, all of which will be needed when germination takes place. Open up a sample seed if you want to take a look. See that the embryo is there and that it has a good food supply in the cotyledons or endosperm. Save all those seeds which look heavy and plump enough to be complete. As for the little, weak seeds, they probably won't survive a storage period. If they do, they may not germinate. Even if they do germinate, they will grow into weak, spindly seedlings, which you'll weed out anyway. So toss them out at once and don't bother to store them. Make sure that the seeds you do retain are absolutely clean. You don't want bits of fruit, chaff, humus, dirt, or any dead matter to rot around your live seeds.

Providing Dry Storage Good, dry seeds should go into good, dry containers. But there are two schools of thought about whether these should allow air to get in, or be airtight. If the seeds are to be stored dry and you can keep the humidity low, it may be a good idea to use loosely closed containers. Living seeds obviously need some air for respiration, even though it continues on a very low level during storage. Modern methods have been developed, however, in which the seeds are chilled at a low humidity and then immediately enclosed in a sealed packet. Evidently the tiny bit of air in the sealed packet is sufficient for the seeds, especially when the packet is in cold storage.

For most of the seeds you store dry, you can use envelopes of folded paper with the ends somewhat open. (Paper is preferable to polyethylene since it can absorb any moisture.) If the seeds are not totally dry when you are ready to put them in the envelopes, spread them on paper on a dry shelf or sunny windowsill for a few days, whether they are still in their fruits or not. If it is too much trouble to make all those envelopes, use little boxes or store-bought envelopes. If moisture begins to gather at any time, put the seeds in a nylon stocking and press the "cool" button on a hair dryer to air them.

Or you can rig up a box with an electric light bulb in the bottom to supply a little heat to dry out seeds. The light is really better because it provides a steadier heat, and the evaporation goes on without interruption as the heat rises. A 25-watt bulb should be modest enough to prevent overheating. If the heat goes above 65°F., however, turn off the dryer or bulb. If you hang seeds in paper bags, put them as near the ceiling as possible, and tear a corner off the bag so moisture can escape if any happens to accumulate. Never use a box or tin that will collect moisture which just drops back down on the seeds. The best arrangement is one that provides slow, gradual drying and no return to moisture.

Here seeds saved from a home vegetable garden have been put in cloth bags and are being gently dried in a cardboard box fitted with a light bulb.

Drying agents such as the deliquescent capsules put into pill bottles can help to keep the moisture reduced around your seeds. So will a packet of 2 teaspoons of freshly opened dry milk wrapped and tied in several layers of facial tissue.

For ordinary storage, usually thought of as dry storage, the ratio of temperature to humidity is important, too. Some seeds must not be allowed to get too dry. If allowed to dry out, wheat seed, for example, will split at the end opposite the embryo. This brings on a condition which will make it impossible for the seed to take up water and start gas exchanges when time for germination comes. The seed will die. So containers for wheat seeds should protect them from overdrying. Cold storage at the proper humidity level will help.

Carrot seeds will do well for a year in temperatures as warm as 80°F. if the humidity is 7 percent. If the temperature goes down to 70°F., the safe humidity level is 9 percent, but in cold storage (40°F.) carrot seed can survive with a humidity of 13 percent.

Onion and tomato seeds need about the same temperature/humidity ratio as carrot seeds, but lettuce and peanuts, for example, need air considerably drier. (For most vegetable seeds stored at room temperature, humidity should be between 5 and 7 percent. Around 8 percent is okay for beans, sweet corn, and spinach seeds.)

If the 13 percent humidity tolerable to carrot, onion, tomato, and some other vegetable seeds at 40°F. is present at higher temperatures, fungi will become active; and at 8 to 9 percent humidity at the higher temperatures, insects will get active, too. And when they reproduce, trouble begins. When humidity is above 18 or 20 percent, the seeds may heat up, and that is bad for them. At about 40 to 60 percent humidity, there is enough water intake for the seeds to begin to germinate. These are all conditions for the gardener who saves his or her own seeds to watch out for.

Longevity and Viability

Seeds stay viable for varying lengths of time. In general, for each decrease in humidity of 9 percent, the life of the seeds tends to be doubled. If you are storing seeds that do not have to be kept moist, the good general principle is to store a well-dried set of seeds in a tight container in a cool place. This is especially advisable if you have a fluctuating and more or less unreliable environment for storing seeds. Evidently the respiration needs of seeds are reduced when the temperature is really cool (as during winter storage outdoors.)

If you provide these favorable conditions the life expectancy of seeds is considerably lengthened. Some seeds are more likely to last well while cold than others. This group includes barberry, ceanothus, cypress, huckleberry, osage orange, apple, mountain ash, redwood, grape, snowberry, and a good many needle evergreens such as juniper, pine, hemlock, and spruce. In fact, certain seeds that are viable for 5 years in traditional open storage are now known to stay viable for considerably longer periods in cold storage. The 5-year group included cucumber, endive, watermelon, muskmelon, and such flowers as African daisy (*Arctotis grandis*), Shasta daisy (*Chrysanthemum leucanthemum pin-*

natifidum), sweet pea (*Lathyrus odoratus*), and marigolds, nasturtiums, zinnias, pansies, and petunias, and several others. Because of their very hard seed coats, some of the bush and tree seeds have been thought of as being viable for 10 or 15 years in ordinary storage: this group included Russian-olive (*Elaeagnus angustifolius*), sugarbush sumac (*Rhus ovata*), black locust (*Robinia pseudoacacia*), various species of eucalyptus, and the Siberian pea-tree (*Caragana arborescens*).

Vegetable plants believed to be viable only 1 year under ordinary conditions included sweet corn, onion, parsley, and parsnip; among the flowers candytuft, summer cypress (*Kochia* spp.), and perennial delphiniums are good for a year in ordinary storage. In fact, it was considered lucky if delphinium seeds lasted as long as a year.

In between are seeds which traditionally lasted 2 years: China aster, straw-flower (*Helichrysum bracteatum*), beets, and peppers. Those said to last 3 years without chilling are asparagus, beans, celery, lettuce, peas, spinach, tomatoes, and quite a few flowers. With cold storage we now know definitely that the viability of several of these can be prolonged quite a bit. For example, asparagus and beans are now known to be able to last up to 8 years; celery, from 8 to 10 years; carrots and lettuce to 9 years; peas and spinach, to 7 years; and tomatoes up to 13 years. Kept cold, the cabbage and squash groups can last 4 years, and pumpkin up to 8 years; eggplant, okra, and the radish and turnip groups for 4 to 10 years. As you may have guessed from your own garden, with no cold storage except the cold ground, purslane lasts up to 20 years or more. Though poplar seeds supposedly need planting within a few days after ripening, seeds of the lotus (*Nelumbo nucifera*) germinated after 237 years in dry storage.

So nature provides us with the full gamut. Try your hand at saving seeds yourself in ordinary or cold storage and see how long they can last. If you are young enough, you may be able to plant the seeds of some evening primroses or smartweed kept in an airtight bottle for 50 years. Success will be more likely if you can keep them in a cool cellar or refrigerator. How long frozen seeds actually will last I do not know, but I have read about finds in the Arctic that make me think it may be a very long time.

The viability times for various seeds given in Part II are averages established by gardeners and seed testers using ordinary storage measures. If you practice cold storage instead, you can extend the viability of many seeds considerably. I'd certainly like to encourage you not to throw out good (or probably good) seeds: if you don't use up a whole packet one year, the leftover seeds may be viable for use the next year, or several years hence. Also, since conditions do not always favor germination, it is good to have such a supply of seeds to fall back on for second plantings. Use your refrigerator.

Dormancy and Special Treatments

Although many gardeners refer to the period between seed maturation and germination as dormancy, it might better be called rest or storage. Most botanists and plant physiologists who study seeds define a dormant seed as one which has the moisture, oxygen, and temperature required for germination but still does not germinate. This situation is rarely encountered among garden or crop plants, for any dormancy mechanisms present in wild ancestors have been "bred out" by human selection techniques. However, many wild plants and woody plants—including some trees and shrubs used by gardeners—do have some type of dormancy.

Dormancy has various causes and can be broken by a variety of treatments. One type of dormancy is caused by seed coats impermeable to water, air, or both. Obviously, before germination can occur, the seed must take in water. In nature, the seed coats are broken or weakened in several ways. As I mentioned earlier, they can be rubbed against rocks. Or they can pass through an animal's digestive system. Propagators sometimes weaken hard seed coats by soaking the seed for a few hours; this is done with morning glories, for example. Though you can usually use warm water, a few seeds may need to have boiling water poured over them and then be allowed to cool. Indigoes, lupines, and members of the pea family need soaking longer, for 12 to 24 hours. However, you should pour the water on and off occasionally to give the seeds access to oxygen. When seeds are so treated and put in a cool dark place, such as a room with a green night-light, the combined methods are called *vernalization*.

Another method involves switching seeds back and forth between the freezing compartment and the lower part of a refrigerator so the seeds go from 40° to about 28°F. By moving them back and forth quite a few times, you can imitate the alternate freezing and thawing they would undergo in nature. (Use a medium with a moisture content of about 10 percent.) A third treatment sometimes used calls for soaking seeds for 20 minutes in sulfuric acid, then rinsing them with water for 3 hours. I don't recommend this, and I am not going to say a word more about it.

Propagators often use *scarification* to break or rupture seed coats. With this method, you nick the seeds with a knife, shake them with sharp pebbles, rub them on sandpaper, abrade them with a file, or put them through a blender, being careful to include some water. (If your metal blender blades are too sharp for this purpose, you can replace them with hard rubber ones, that is, if you're going to process enough seeds to make such an operation worthwhile). You can scarify the coats of larger seeds or pits of stone fruits individually by cracking them in a vise, but be careful or you may destroy the embryo.

Some seeds contain inhibitors which prevent germination. These may reside in the fruit around a seed, in the seed coat, in the endosperm right next to the embryo, or in the embryo itself. In very dry climates, for example, some seeds will not germinate until the rains come and release the inhibitors. As this fact suggests, in some cases you can wash inhibitors out of seeds by using flowing water or by soaking the seeds and changing the water frequently. Sometimes you can destroy or change inhibitors through a moist, cold pretreatment, which is explained below.

Stratification   Herbaceous plants, many woody plants and cultivated garden flowers, and some vegetables are characterized by a gentle internal dormancy, probably a type of embryo dormancy. Fresh seeds of such species are not ready to germinate until they have dried out and had a resting period while dry. This necessary after-ripening may last anywhere from a few days to a few months. Usually people are quite content to wait until the seeds are ready. However, propagators in a hurry sometimes manipulate light and heat or use gibberellic acid to break this kind of dormancy.

The most common pretreatment is prechilling, or *stratification,* a term derived from the old method of putting the seeds down in layers, or strata, of wet sand. Today, though, the usual method is to put the moist seeds in a well-closed polyethylene bag and place this in the refrigerator or cold cellar at 40°F. The seeds must be wet, because the changes that take place during stratification can occur only if the seeds have imbibed water and been rehydrated. You can provide this moisture by adding water or packing the seeds in moist sphagnum moss or sand. Do make sure that the stratification temperature is not too low: temperatures below freezing can kill seeds which have already been rehydrated because ice crystals form in the cells and rupture them. Since stratification duplicates the conditions of winter in nature, you may want to plant seeds in the fall for germination in spring. Different kinds of seeds have varying stratification requirements; some examples are 1 to 3 months for apples; 2 to 3 months for firs; 2 months for false cypress; and 3 months for dogwoods and mountain ash.

Multiple Dormancy Requirements   Many plant seeds have a single dormancy mechanism requiring only one type of treatment. Some seeds, however, are controlled by more than one dormancy mechanism and need several special measures. Blackberry seeds, for instance, require scarification and then 3 months of warm stratification followed by a similar period of cold stratification. Just about the same pretreatments are necessary for raspberry and dogwood seeds. Of course, these seeds can be scarified and then planted in the early summer so they receive the warm and cold pretreatments naturally. (Actually, most people who grow bramble fruits prefer asexual methods like layering and bother with seeds only when they're interested in breeding new varieties.)

Seeds like those of the northern viburnum respond to warm stratification with emergence of the radicle, or seed root, followed by cold stratification with emergence of the epicotyl, or shoot. Lily seeds may behave this way too.

It is always a good feeling to be able to propagate wild flowers from seeds because then you know that you are not going to disturb the original plant, even though some of those which you may want to introduce into your own garden are very common and very rugged. Of those which are considered possible to grow from seed, a few demand special treatment. Most, however, can be grown from seeds harvested when the pods are ripe. To be on the safe side, it is a good idea to put a little bag over the pod, in case it is going to burst and send the seeds flying before you have a chance to collect them.

Many wild flower plants are capable of being propagated by dividing rhizomes, offsets, or underground stems, while others can be increased quite easily by cuttings. I believe in providing this information about the possibilities of growing wild flowers from seeds, however, because in many cases it may be only prudent to attempt propagation this way. You may not want to go digging around in someone's back lot, though it may not be difficult or imprudent at all to walk through and gather up some seeds someday.

Professionals warn against certain attempts with difficult seeds because of the probability that the amateur grower will not have the right habitat to provide, once the wild flower plants are up. It is true that some need mountain air and temperatures, that others need wet spots, and that still others need cool, acid, moist woods.

Most important of all is the acidity which quite a few wild flowers need in order to survive. A very acid pH of 4 to 5 is required by such wild plants as bearberry, bog violet, buckbean, checkerberry, devil's bit, dwarf cornel, featherbells, fringed polygala, meadow beauty, oconee-bells, painted trillium, partridgeberry, trailing arbutus, turkeybeard, twinflower, wood lily, and wood sorrel—some of these so rare and so difficult and so dependent on special bacteria and

Working with Wild Flower Seeds

145

fungi that it is just about impossible to create suitable habitats for them without extraordinary means to keep out all nonacid soil influences or soil inhabitants. (What's more, some of the more delicate wild plants are great favorites of mice and slugs and snails, and as seedlings and small specimens they disappear in no time almost as soon as you set them out.)

Certain wild flower seeds need quite acid soil of pH 5 to 6. These include the anemones, baneberry, clintonia, dogtooth violet (trout or faun lily), foamflower, galax, gay wings, green dragon, hepatica or liverwort, mayapple, mitrewort, purple trillium, tickseed, and wild geranium.

Those which are able to exist in soil of p.1 6 to 7, or a soil which is more or less like the usual garden soil except in limey or arid areas include, for woods flowers: alumroot, bloodroot, blue cohosh, claytonia, columbine, confederate violet, Culver's root, Dutchman's breeches and squirrel corn, jack-in-the-pulpit, Jacob's ladder, *Phlox divaricata*, Rocky Mountain columbine, showy lady's slipper, Solomon's seal and false Solomon's seal, Virginia bluebells, wild ginger, wild hyacinth, and the only lady's slipper which I will recommend gardeners to attempt, the yellow lady's slipper. There are others, but these are enough to give you an idea of the problems. Be sure your attempts fall within the realm of possibility as far as your own land goes.

Another factor that discourages propagators interested in wild flowers is the warning that you cannot grow certain seeds unless they are dowsed with fungicides and other such poisons.

Instead of giving up any attempts, try instead, a fairly simple and safe way of planting seeds in sterile conditions. Pack moist peat and sand in a 3-1 mixture into a pint mason jar up to within 1 inch of the top and cover the mixture with ¼-inch of sand. It is best to use coarse mason's sand. Make the mixture more moist until it is a little wetter than a wrung-out sponge, and put on the cover, loosely. Then set it in a 160°F. oven for 60 minutes. Leave the jar covered after you take it out of the oven to cool. Wash your hands with Dial® soap or rub them with alcohol and sow the seed, replacing the cover immediately afterwards. Place on a bright windowsill which is shaded from direct sun, at a temperature of about 68°F. (a good average for wild flowers). The seeds which require a cold pretreatment, or those whose directions include a statement about letting the seeds lie in the cold or the cold frame for the winter, may be planted in jars and then transferred to the refrigerator at about 40°F. until the days lengthen in the spring and it is time to bring them out to grow.

When you put them on the windowsill right away, or after removing jars of seeds from the refrigerator, it is best to remove the cover once a day so that the germinating seeds and little sprouting plants can get some air. Do it for just an instant the first couple of days, and then for a minute and then a few minutes. When the little plants push up against the lid of the jar, they are probably big enough to transplant. At any sign of damping-off fungus, which there should not be, of course, remove the lids at once and clean out the growth in hopes of saving your plants. Once they are established, and once up and ready to be transplanted, transfer to a sterilized peat pot will help to prevent a later transplant shock when the little wild flower plants are put out in the ground. This is especially necessary for plants with taproots or especially delicate roots. Under ideal conditions the seeds of such plants can be planted directly outdoors, but ideal conditions are not always easy to get. The only consolation is that damping-off is a disease much rarer with plants grown outdoors than with those started under artificial conditions in the house or greenhouse.

Delicate plants which should not be moved when they are little like a starting mixture which has some soil and nutrients in it to support the seedlings once they are up. A 1-1-1 mixture of peat, sand, and soil is best. This soil can be steamed to pasteurize it or it can be baked in a 160°F. oven for an hour. Top this mixture with sifted sphagnum moss. If the seeds are large, plant them and cover them with more sphagnum moss. If they are small, simply drop them onto the moss and force the seeds down into the moss with a quick, strong spray of water from a bulb spray. Afterwards water only from below. Any covering used to keep the humidity high should be removed regularly to let in air. These precautions are necessary to protect the plants from disease. And the longer you can support tender wild flowers in places where there are no slugs and mice, the better.

As you probably know, many states now have conservation lists informing the public which plants are endangered and on the protected list. Get the list for your state, and consult trained botanists or people at the state university to see what they would advise, and what extra instructions they can give you about your own state's needs. You can also get a copy of the Smithsonian Institution's list of endangered species for the whole country. Then if you find something you can easily propagate and put back into its habitat in your area, you will get a good bit of satisfaction when you succeed. If, of course, your state specialists advise otherwise, it's best to let well enough alone. (See Part II for some suggestions on wild flowers to try.)

Planting and Germination

Many books and government pamphlets have been written on how to plant seeds, and seed companies often put excellent directions right on their packets. Because of this, I'll emphasize only a few points.

What Happens in Germination

As I said earlier, a seed has very little water content. This dryness is undoubtedly the reason why seeds can remain so long in a state of suspended animation. Before germination can occur, the uptake of water (imbibition) must occur. During imbibition, water molecules rehydrate all of the molecules of the seed. The seed swells, and the enzymes are reactivated and begin to do their work.

Food storage molecules in seeds are usually large complex ones such as starches, fats, and proteins. With the help of enzymes, these are broken down into smaller units: sugars, fatty acids, and amino acids. The smaller units are used in two ways. Some of them are consumed in respiration, producing the energy needed for the rapidly growing embryo. Others are re-formed into larger molecules, such as cellulose, which are the building blocks for new cells. As respiration increases, the uptake of oxygen must also increase.

Usually the tiny root (radicle) is the first structure to emerge from the seed. It serves to absorb water and minerals for the rapidly dividing cells. The first structure to push up through the soil varies with the plant. In some, such as corn and peas, the cotyledon or cotyledons remain in the ground; in others, like lima beans, the cotyledons emerge. The new shoot tip—the epicotyl of the embryo—produces the first true leaves. These young green structures begin to photosynthesize and the new plant is on its own.

Planting Seeds

Everything you do in planting seeds is governed by the seeds' biological needs during and just after germination. Many a seed has died because of an insufficient supply of oxygen, for respiration is very fast at the start. This is why experts advise you to have nice loose soil with excellent drainage to get rid of excess moisture and why seed packets and every other set of instructions caution you not to plant seeds too deep. The usual advice is to plant at a depth of 3 times the diameter of the seed. I usually plant at this depth, though I also have seen 2 and 4 times the seed diameter recommended. Weather conditions are important. If the season is very wet, do not plant too deep.

Seeds That Need Light

Some seeds, however, should always be planted much more shallowly (preferably under just a thin layer of loose vermiculite) since they need light in order to germinate. This group includes:

African Violets (Saintpaulia *spp.*)
Ageratum
Aralia, the house plant
Begonias
Bells-of-Ireland (Molucella laevis)
Browallia
Burning Bush (Kochia scoparia childsii)
Chamomile or Feverfew (Matricaria *spp.*)
Christmas or Jerusalem-Cherry (Solanum pseudocapsicum)
Cineraria
Cloud Bent Grass (Agrostis nebulosa)
Coleus

Columbine (Aquilegia *spp.*)
Coral-Bells (Heuchera *spp.*)
Creeping Zinnia (Sanvitalia *spp.*)
Daisy (Gerbera *spp.*)
Daisy (Venidium *spp.*)
Dusty Miller (Centaurea cineraria)
Edelweiss (Leontopodium *spp.*)
Endive
Figs (Ficus *spp.*)
Firecracker or Cigarflower (Cuphea *spp.*)
Flowering Tobacco (Nicotiana *spp.*)
Foxglove (Digitalis *spp.*)
Fuchsia
Gloxinia
Golden-Cup (Hunnemannia fumarifolia)
Grevillea
Impatiens
Jacobinia

Kalanchoe
Lettuce
Lobelias
Maltese Cross (Lychnis chalcedonica)
Naegelia or Smithiantha Zebrina
Penstemons
Petunias
Portulaca
Primulas
Queen Anne's Lace
Rock-Cress (Arabis *spp.*)
Scarlet Sage (Salvia *spp.*)
Sinningias, the house plant
Slipperwort (Calceolaria *spp.*)
Snapdragon (Antirrhinum *spp.*)
Spiderflower (Cleome *spp.*)
Stocks (Matthiola incana)
Straw-flower (Helichrysum *spp.*)
Sweet Rocket (Hesperis *spp.*)
Yarrow (Achillea millefolium)

Factors Affecting Germination and Growth

In respiration, carbon dioxide is given off, and if the seeds are planted too deep for this gas to escape into the air, it may delay germination and keep the seeds rather dormant. As it escapes, the carbon dioxide may help to prevent crusting of the soil surface. This condition is detrimental to the entry of necessary oxygen, though undoubtedly not so detrimental as too much water in the seedbed, which really blocks the entry of oxygen.

But some water definitely is needed. That is why the experts tell you to have humus, compost, peat moss, vermiculite, or other water-holding materials in your soil. It is better to have the water embedded in these materials than running loose and clogging up the spaces between soil particles. The need for water also makes it wise to shade little plants from hot sun and drying winds. Otherwise they will dry out and die. A mulch can help too.

Temperature is another factor you must take into account. The seeds of some plants germinate within a rather narrow temperature range. Celery, lettuce, and onions require cool soil for seed germination, that is, a soil temperature of not more than 68°F. Cucumbers, muskmelons, and okra, on the other hand, need a warm environment of above 68°F. for germination. Other tropical and semitropical plants also must have higher temperatures when getting started. Asparagus, tomato, and sweet corn seeds need at least 50°F. to germinate, and squash and some bean seeds require at least 60°F. The embryos of soybeans, lima beans, cotton, and sorghum will be injured if cold below 60°F. strikes them when the seeds are first taking up water. Be careful, however, not to heat seeds above 80°F., or at the most, 95°F. Also, make sure you accompany any such heat during initial germination with somewhat lower temperatures at night. When the plant structures actually begin to grow, go to lower temperatures during the day. While using higher temperatures, remember to increase the light and day length and watch closely for diseases. (If in doubt, you can add extra carbon dioxide as described under The Right Environment for Propagation in chapter 1.)

Of course, with vegetable and other plants you start outdoors you won't have to worry much about temperature. Usually you'll choose to plant seeds of species you know are suited to your zone, and you'll put them in at the recommended time since they will grow best then. Don't rush to plant, for seeds often will not grow in cold, soggy soil not yet warmed by the spring sun.

Needless to say, light is needed by all plants from the moment that the first leaves appear above ground and photosynthesis begins. Direct sunlight does no harm as long as the plants are protected from heat and drying out, which do them a lot of harm. If you plant your seeds in a flat in moistened sphagnum moss or vermiculite and water from below, you'll have a better chance of keeping small new plants from drying out. As I suggested above, your best practice is to reduce the warmth (and any bottom heat) as soon as germination has taken place. Unless the plants have to have heat, lower the temperature to 55° to 60°F. after germination. With this reduced temperature and plenty of light, you should have stocky, strong plants.

Spacing is also important to your plants. Tiny plants need to grow close enough to benefit each other; they surely do this, though no one knows yet whether the mutual aid comes from conditions and interchanges occurring underground or just aboveground (shared nutrients, dew, shade, or whatever). Yet seedlings must not be so near together that the warmth and lack of airiness encourage damping-off, the fungus pest I discussed in chapter 1. To deal with this paradox, take the following steps. First, tap the seeds out of a folded paper—not the packet—when you plant them, and try not to let them fall too close together. As soon as the seedlings are up, you can begin to snip off the weakest with nail scissors to give the rest a little more air. Finally, when the seedlings have 2 to 4 true leaves, you can lift up the little plants with a forked stick, separate them very carefully and replant. This time make the distance between them about 1 inch, far enough so their leaves just miss touching.

As they grow, replant again, and again, so the leaves just fail to touch. You can pinch off center buds to encourage side-branching. For plants which can stand it, undertake several replantings. Disturbed roots often grow bushy and strong, and the moving does them more good than harm. (Exceptions include the taprooted legumes, persimmons, and papaws.) At all times, make sure your seedlings get adequate light, and get them outdoors as soon as possible.

Fertilizing   Be very careful about giving fertilizer to very young plants, except as necessary to supply nutrients to plants started in sterile mediums. Too much nitrogen, for example, will create a spindly, succulent plant that is weak in the wind and irresistible to insects. It may never recover from this early weakness. You're wiser to feed only enough nutrients to keep your young plants constantly, slowly growing; mild weekly feedings are often enough.

Seedlings grown indoors usually are hardened before they are transplanted into the garden. Many gardeners think of this process as acclimating the plants to cold or to the heat of the sun at midday. Actually what such exposures do is check the growth of the plant. Then carbohydrates accumulate, and with these resources the small plant is better able to withstand the shock of new conditions when taken up and put in the open garden. Beware, however, of plants you buy in flats, which may have been checked too severely. If they have suffered a real setback, they will never be able to catch up to a healthy, well-managed plant.

Always water plants you are about to move. And keep as much of the soil they grew in as you can. If you are putting out small plants in peat pots, crush the pot with your hand. Tear off any rim which protrudes above the soil line, and be sure to bury all of the pot. These precautions will help to prevent the peat from drying out like a wick and injuring the tender small roots. They will also help the pot to rot faster in the soil.

If you have used net-covered peat pellets, it is advisable to remove the net. Though the directions say this is not necessary, I have found that the net does inhibit root growth and prevents the intermixture of soil which helps a plant make an adjustment. If long roots have been constricted and deformed in those pots, you should remove the pots carefully and do some judicious root pruning. In any case, root pruning of small shrubs and trees is desirable. Grow them in screen-bottomed flats, and snip off any roots that have emerged about 1 inch when you transplant.

When you position transplants, remember that all plants which do not have a rosette growth pattern may be placed deeper in the earth than they were before transplanting. But, rosette plants like petunias or primulas must never be set with their crowns below the soil surface.

Good soil that is well fertilized with compost, together with a cooling, moisture-holding mulch, will help transplanted plants take hold and flourish. If the sun is hot, shield them with brush or caps, or make a lath structure to put on for partial shade when needed. (See chapter 1.) Water and mist will help a lot if the plants seem droopy. Keep watch, for small plants have a trying time just after they are transplanted.

But once established, they are on their way to make flowers, more seeds, and start the cycle again.

PART II
Plants to Propagate

A Preview and
Some Suggestions

One of the enticing things about propagating plants is that you can experiment with breeding them yourself. The listings which follow give information about how to apply preferred methods of propagation to a large number of plants. You'll also find a few hints on culture and some details about breeding techniques.

The first listing—of fruits, berries, and nuts—focuses especially on breeding possibilities because this group of plants offers what seem to me the most exciting results for the amateur who wants to experiment.

In the next listing—ornamental trees and shrubs—you'll find information on both sexual and the vegetative methods of propagation which are more commonly used for this category of plants. Sexual methods are difficult for woody plants because they involve complicated dealings with dormancies of seeds. But details on how to overcome these very precise protective devices of seeds are given for those trees and shrubs that you have a good chance of growing successfully from seed. Try asexual methods when possible.

Since herbaceous plants usually die down over the winter, their seeds have not had to develop unusual protective measures against weather. Therefore, the listings which follow on garden flowers, wild flowers, ferns, vegetables, herbs, and house plants do not stress so many of the special requirements of seeds (or spores) to be met for successful propagation.

Important for all plants is a suitable soil environment: the appropriate climate zone, the right soil with right pH, moisture, air, nutrients, shade and—for some acid-loving and other plants—mycorrhizal associations with soil organisms vitally important for certain wild plants. The sterile conditions that propagators can provide usually compensate for the lack of control we have over natural predators and for the antibiotics common in the outdoor environments where plants live. Some special needs are noted in the plant listings, but many basics are skipped because such cultural requirements are commonly given in other books about raising plants, and are really outside the province of the main topic of this book.

Ingredients and characteristics of typical potting mixtures are given in Part I. (See chapters 1 and 2.) As I mentioned there, many, many mixtures for seeds, cuttings, and bulbs have proven successful in the past. Remember to meet the particular acidity (or neutrality or alkalinity) needed by the plant you are working with; these pH preferences are given in most books dealing with the culture of plants. Also provide for the plants' requirements for porosity and a water-holding capacity in the medium, and meet the need for sterility and cleanliness in mixtures deprived of natural controls. If you find that none of your seeds of a certain plant or none of your bulbs and the seedlings succeed, suspect rodents. Repellent herbs or bulbs, cages around your bulbs, seeds, or seedlings, and Hav-a-hart® traps can help you out. Even an environment like a dark-colored bottle with some water and bits of charcoal in it is adequate to induce new growth in some stems or leaf cuttings you may want to root.

I have a few more admonitions: when taking cuttings from underground parts of a plant, always try to know which end should go down, which up. If in doubt, keep track of which end has been proximal to the plant, and if in double doubt you can always plant the sections horizontally and let the plant decide. With stem cuttings (need I add?) that last resort is not so likely to be successful. (See chapter 2.)

The question of whether or not to use bottom heat is a rather tricky one. It is true that quite a few cuttings and seeds do succeed better if you can start them with heat from below, provided either by a hotbed or from a cable in the soil of the flat. Yet many plants seem to grow just as well without this help. Another tricky question is whether or not to use growth-promoting agents. Again, many plants seem to grow just as well without them. If you want to use both these helps, I see no good reason against it because both are very like what happens in plants naturally (when they wait for warmth in the spring or wait for the flow of hormones which trigger new growth in the proper season).

Misting is even more important to consider. A fine spray of mist imitates fog, rain, and dew, which certainly need replacement in indoor habitats. Also the wetness you put into the air from mist-sprays is a great balm and aid to a wounded plant or piece of plant under pressure of the unfamiliar conditions you impose on it during propagation.

The vegetative methods of propagating and improving plants are relatively fast compared to the generation-after-generation breeding needed for improving by seed. Unswerving genetic purity may result from vegetative methods; but maybe you could come up with a pink marigold or a new tree bean if you persist in your work of saving and planting the most likely seeds.

It is impossible for lists like those which follow to be complete, or anywhere near it. I hope, however, that quite a few of your favorite plants are included and that you find the hints you need for the specific plants you want to propagate.

Fruits, Small Fruits, Wild Fruits and Nuts

Almond (*Prunus amygdalus*). The best method is bud grafting on seedling rootstock. Use scions of 'Nonpareil', the old favorite; or 'Bartre' with large kernels; or 'Texas' or 'Mission' with thicker kernels than 'Bartre'. These scions are sometimes incompatible with some stocks. For stocks try other almonds, especially bitter almonds, or try peach, plum, or peach x almond hybrids. (Especially useful is peach stock 'Lovell', but good also is 'Nemaguard', both valuable for irrigated areas because their roots are more tolerant to water than are almonds'.) Resistant to oak fungus are plum rootstocks such as 'Marianna 2624', though incompatible with 'Nonpareil'. Use instead stock of 'Mission' and its progeny. F_1 crosses of 'Nemaguard' peach and 'Titan' almond make good vigorous stock material.

In order to grow almonds from seed you must cross-pollinate because almonds are self-incompatible. Follow directions for collecting pollen, emasculating, and pollinating given in chapter 6. Collect pollen early, and force in warm room if necessary. When seeds are formed, soak if dry for 12 to 24 hours before stratifying at 40°F., or plant in fall in cages outdoors. Germination is best at no warmer than 68°F. or seedling abnormality may result; if it does, pinch back and you may get new normal growth.

Breeding aims should include working for self-fertility, hardiness, and varieties that are not subject to noninfectious bud failure. Almonds have taproots. Be careful.

American highbush cranberry (*Viburnum trilobum*). This relative of the elderberry and the European *V. opulus* is a preferred bush to grow and use for breeding and propagation because it attracts no aphids and does not have the crinkled, distorted leaves sometimes seen on *V. opulus*. The fruits are clear, red, and acid. To grow and breed, try one cultivar, 'Wentworth' for early fruit, one midseason, 'Hahs', and one late, 'Andrews'.

For extra hardiness, try the Canadian clone 'Manitou' or a very pleasant tasting one, 'Phillips'. Other viburnums which may be crossed with *V. trilobum* include *V. edule*, *V. flavum*, *V. kansuene*, and *V. orientale*.

Apple (*Malus* spp.). Most apple propagation, vegetative or by cross-pollination, by hand (because apple is self-incompatible). *Malus pumila*, the common apple, more hardy when crossed with *M. baccata;* or with *M. floribunda*, the Japanese crab apple, also used for disease resistance. Virus resistance probable in *M. sieboldii* and *M. sikkimensis*. Cold hardiness in rootstocks of *M. coronaria* 'Nieuwlandiana' and *M. robusta*. For dwarfing use 'East Malling IX'; also called 'Malling IX'. For semi-dwarfing 'MM III' or 'MM106', which does not sucker.

Apples, like many woody plants, have juvenile period (leaves small, shoots thin, no flowering) and another transition period when lower part of plant is juvenile, and upper, mature. For cuttings, use juvenile growth. (Mature won't root.) In grafting match the phases. You can shorten the 2- to 10-year juvenile period by grafting buds of seedlings onto *dwarfing* rootstocks. For budding and grafting, rootstocks and scions should be compatible, in good health, free from viruses. Use stocks without suckers, able to produce heavy crops, and to grow into big trees with good vegetation, not the more dwarfed trees which weaker stocks will produce (as will root pruning, pruning, ringing the bark, or a too-heavy crop of fruit.) Scions on dwarfing stocks of 'McIntosh' and 'Cortland' tend to be susceptible to scab fungus, but 'Golden Delicious', 'Baldwin', 'Wealthy', and 'Rhode Island Greening' are not. Do not plant graft union below ground level, especially for dwarfing stock. Several varieties of scions on one stock will be less sturdy than trees with a single graft; prune to keep one strong scion from taking over. Stake weak trees and wrap to prevent sunburn and rabbit nibbles.

To propagate by seed, hand-pollinate so as to over-come incompatibilities and help along bees, which cannot always transfer pollen from long stamens to short styles in apple blossoms. Collect flowers at balloon stage, rub off anthers, leave in warm place to dry and let pollen fall. Use immediately on emas-culated flowers or store dry in refrigerator, using calcium chloride if necessary to keep the pollen dry. If necessary to keep for a year, put in deep freeze. To emasculate flowers, follow directions in chapter 6, doing 2 flowers per cluster. Use bags to keep wounded flower parts from drying out. Pollinate with camel's-hair brush or finger.

Harvest seeds when fruit is ripe, stratifying in moist peat moss at 35° to 40°F. for 6 to 14 weeks. Then hold seeds at 60°F. until time to raise the temperature to cause germination. Weak Clorox® rinse helps to keep down fungi.

To breed for yearly bearing, choose 'Cortland' and 'McIntosh'. For alternate years: 'Baldwin', 'Golden Delicious', and 'Wealthy'. Try out some old-time favorites: 'Strawberry Apple', 'Chenango', the snow apple 'Fameuse', the small, good 'Lady', the good fall apple 'Porter', and also 'Cox's Orange Pippin', 'Red Gravenstein', 'Pound Sweet', and 'Pumpkin Sweet'. Ask at large experimental orchards about available scions for grafting.

Apricots (*Prunus armeniaca*). Early-blooming and tender (29°F. freeze will kill young fruits in early spring), these trees like dry air, but moist soil. For grafting, use apricot understock or perhaps peach or myrobalan plum. Species of apricot to combine include *P. a. manshurica, P. brigantiaca, P. a. ansu, P. mume,* or the black apricot, *P. dasycarpa* (*P. armeniaca* × *P. cerasifera*), which is both cold-resistant and disease-resistant.

Apricot flowers are usually self-compatible but occa-sionally are self-incompatible. Collect blooms for pollen before the balloon stage, preferably 2 or 3 hours after sunrise. Cool at once, in traveling ice box if necessary. After 24 hours, push through sieve. To collect blooms earlier, cut branches, and bring indoors a month before usual blooming time. When buds swell, cut stems again under water every day or so to guarantee efficient water uptake. When pollen ripens, pick blossoms three times a day. On outdoor trees, emasculate blossoms as late as practicable before flowers open to expose stigma, but not so early pistils will fail to develop properly. Be very careful, for apricots are very tender and fragile. If there's danger of frost, cover emasculated areas with opaque material. Pollinate the day after emasculation, unless it's rainy: if rainy, do it right after the rain. Pollinate twice, for better results. Guard trees from squirrels, and when fruit is ripe, leave on tree up to just before onset of rotting. Then harvest. Put late-ripening seeds in cold stratification for 2 to 4 months, but keep early-ripening ones on agar in sucrose solution unless you were able to get them into stratification before drying began. You can use hypodermic needle to put moisture into seed to keep it moist. Unless already started in the fruit, germination will begin in 2 to 3 months. Plant then; or if seeds were stratified at time of early harvest, plant in late summer in 1-1-1 sand, peat moss, and leaf mold. The chief aim of breeding is the extension of hardiness zones. This goal can be approached by developing buds resistant to winterkill, 1) during dormancy in mid-winter, or 2) during alternate thaws and freezes in late winter, or 3) just before or during blossoming. Cultivars with long bud dormancy are more resistant to ups and downs of temperature than the very sensitive apricots with short dormancy periods. In fluctuating climates, therefore, try 'Zard', 'Central Asian No. 35', and 'Oranzhevokrasny' from the Nkitsky Botanic Garden. Another need is to develop apricots for warmer areas which will not abort buds in warm winter weather.

For budding and grafting, see chapter 5.

Avocado (*Persea americana*). I grow avocados in my Vermont kitchen all the time, and of course have never seen one in bloom and have never picked any fruit. If you live in the right climate for avocado trees, you have your choice of varieties with fruits that stay on the tree for long after they seem to be ripe; avocados ripen only after they are picked. In California, Mexican and Guatemalan varieties and crosses between the two are grown, with 'Fuerte' and 'Hass' often chosen. In Florida crosses between Guatemalan and West Indies races are often grown. In the tropics, only the West Indian. Select West Indian for the lowlands, Guatemalan for 3,000 to 5,000 feet; and Mexican on the highlands. Since seedlings set fruit belatedly, if ever, propagation is usually by vegetative means. Cultivars to set from California include 'Nable', 'Benik', and 'Panchoy' of the Guatemalan race. Graft on a rootstock which can support the heavy bearing of a mature tree. Dwarfing rootstock is advantageous and, of course, helps resistance to disease and to excess salinity.

Pollination can be of two sorts: either the female part of the flower is receptive in the morning and the male is ripe the following afternoon (with the flower closed from noon the first day to noon the second day); or the male ripe in the morning and the female receptive the next afternoon. The bloom comes only in fair weather when the day temperature is just about 80°F. and the night temperature 58° or 60°F. Sometimes when cold weather delays the male pollen, the pistillate part of the flower will still be receptive even though the prime period failed. An avocado tree bears hundreds or even thousands of flowers and up to a million on a big mature tree. Professionals believe that a mere 1 percent ever come to fruit. Trees (even in tents) rarely set fruit, so you must introduce bees and water into the tent, which preferably has two trees in it. If you try to cross-pollinate, do it in alternate years, which are the 'on' years for fruiting. Only try a few flowers in each cluster, and remove the others before you apply pollen. Be very careful, avoid any dark stigmas, and work during the first, not the second opening of the flowers.

Germination takes place readily, if the seed is viable, sometimes even when the seed is still in the fruit. At room temperature, it will stay viable for a few days after removal from the fruit. Or it can be stored in a humid place at 35° to 40°F. A 24-hour chilling and removal of the seed coat tend to raise the possibilities of germination. When you plant, put the flat end of the seed down, and plant so the top of the seed will be just level with the surface of the planting medium. (See also entry in House Plants.) Keep always moist and at 78° to 86°F. (It will take about a month for germination.) During the next 3 months, if not chilled and not retarded by lack of water or nutrients, the tree will grow 3 feet.

Budding of young seedlings is often done with 'Fuerte', 'Benik', 'Nabal', and 'Itsamna'. The rootstock should be rot-resistant. Crosses are now being sought to overcome rot. One sturdy clone is 'Bacon' and another is 'Zutano'. Early-maturing avocados are 'Irving'; mid-season are 'Rincon' and 'Hass'; and note that 'Hass' is harvested every month of the year in California. 'Fuerte' matures earlier, but stores on the tree for just about as long. 'Creelman' and 'Teague' mature in the fall.

Blackberry (*Rubus nigrobaccus* et al.). Many methods are successful with these bramble plants: softwood cuttings, hardwood cuttings, suckers, root cuttings, and seed. Seed needs treatment for double dormancy: stratification at room temperature for 3 months, then at 40°F. for 3 months. Cuttings and suckers are easy to take. Intermittent mist will help for cuttings. Move suckers in early spring, selecting the most vigorous in autumn to save. Root cuttings in fall (2 to 4 inches long); store in sand at above-freezing temperatures until spring. Natural tip layering is common. Divide plants with sharp spade in early spring.

Flower buds on Eastern varieties begin to form in August; on Oregon varieties in October, though if of

Himalayan origin, not until January. Flowers open over a 2- to 3-day period, 20 to 60 from a few leaf axils. Large anthers easy to remove for collecting pollen, which can be stored at 22°F. with 10 percent humidity. Emasculate flowers when buds are full sized, and cover at once with white paper bag, tied to prevent wind damage. After pollination, replace bag, and replace it later with cheesecloth bag as fruit ripens (to let in light).

When berries ripen, separate seeds from pulp in blender, clean, pour on boiling water, and let sit overnight. Stratify at 68° to 75°F. for 8 weeks, then at 40°F. for the rest of the winter in refrigerator or outdoors. Sow in sphagnum moss. If seeds kept at 70°F., germination in 10 to 15 days. Transplant when second true leaves appear.

Breeding aims include: increased hardiness, large fruit size, seeds that are not too large and hard, good productivity, vigor, insect resistance, thornlessness, and erect and sturdy canes (especially first-year canes). Hardiness adaptation would include long-retained cold resistance, ability to reharden if necessary, and to resist stimulation to growth if temperatures vary. As resource for hardiness try *R. canadensis* and *R. allegheniensis*. Thornless clones 'Smoothstem' and 'Thornfree', unfortunately, have large seeds. 'Austin thornless' less large. Drought-resistant: 'Nessberry'. Anthracnose-resistant: 'Boysen' and 'Young'. Verticillium-resistant: 'Logan' and 'Mammoth'. (Youngberry was developed in 1926; Loganberry in 1881; Boysenberry in 1928.) Locally adapted blackberries include 'Flordagrand' for Florida; 'Flint' and 'Gem' for Georgia; 'Olallie', 'Aurora', and 'Marion' for Oregon; 'Carolina' and 'Williams' for North Carolina; 'Jerseyblack' for New Jersey; 'Raven' for Maryland; and 'Bailey', 'Hedrick', and 'Darrow' for New York. A widely grown blackberry on the West Coast is 'Brainerd'. Also good for West Coast are 'Himalaya' and 'Evergreen'. Other thornless clones include 'Merton Thornless', 'Burbank Thornless', 'Whitford Thornless' and 'Brainerd'. So there are plenty of materials to work with.

Blueberry (*Vaccinium cyanococcus* et al.). There are about 24 species of blueberries plus several huckleberries (*Gaylussacia* spp.), whortleberries (*V. uglinosum* or *V. occidentale*) lin- or ling-berries (*V. vitis-idaea*) and an amazing creeping blueberry which can spread over acres (*V. crassifolium*). Many are self-infertile and insect-pollinated. Many modern cultivars come from F.V. Coville's New Jersey introductions in 1920: examples are 'Pioneer', 'Cabot', and 'Katherine'. The southern rabbiteye blueberry (*V. ashei*) very vigorous, long-lived, sweet, very productive, and heat tolerant. Like others, looks blue before ripe, but is better and darker when truly ripe. Good Northern cultivars include 'Woodard' of very good flavor; 'Bluegem', 'Tifblue', widely used for commerce; and for home gardens 'Garden Blue' and 'Menditoo'.

Propagation from layering, cuttings, and budding and grafting as well as from seeds. Softwood cuttings done under mist-sprays. Evergreen species do better than deciduous species. Easy to root if taken from juvenile wood, midstem, without flower buds. Rhizome cuttings for some species. Blueberries require very acid soil; they have no root hairs and depend on mycorrhizal association with beneficial fungi.

Insects or human hands are needed for pollination because of odd way blueberry pollen grains fall out of the flower. Before stigmas are ripe, pick several branches of 2-year wood to force in water so you'll have pollen ready. (Branches keep best in jar under mist; cut stems every few days under warm water, to retard fungal growth.) When flowers mature, roll between fingers every day to get out pollen. Store in vial or pill capsule, cold and dry, but use fresh if possible.

When stigmas are ready, touch with pollen-dipped sterilized finger at 2-day intervals for as long as the stigma is receptive. When pollen grains germinate, it will take 2 days for pollen tube to grow, and 6 to 8 days for fertilization of egg to occur. When seed matures, embryo is surrounded by a starchy endosperm and a hard, pitted seed coat; about 8 seeds to each fruit. No emasculation needed for self-infertile species, but others should have corolla snipped off with sharp nail scissors. When fruit grows, protect it with nylon net. To save seeds, harvest when dead ripe. Mash seeds from fruit or put through blender for 20 seconds in several short spins. Pulp and skins will float if water added. Germinate without ever letting the seeds dry, or stratify in a moist medium at 40°F. for 2 months followed by alternating cold and warm stratification. Seed germination best if the seeds are left uncovered and have light, and if the day temperature is 80°F. and the night temperature is 50°F. You can plant seeds in flats under red cellophane, with 70°F. day and 50°F. night temperatures. Always use an acid medium. Misting and doses of dilute fish emulsion once a week help.

To get vigor, use cultivars such as 'Tifblue', 'Gordon Blue', 'Homebell', 'Rubel', 'Jersey', 'Stanley', 'Berkeley', 'Blueray', 'Darrow', and 'Murphy'. For very productive bushes select 'Bluecrop', 'Blueray', 'Darrow', 'Tifblue', 'Ivanhoe', and 'Homebell'. Best flavor is believed to be in 'Earliblue', 'Blueray', 'Coville', 'Ivanhoe', 'Tifblue', 'Garden Blue', and 'Stanley'. In the home garden, be sure to plant several varieties together, and to maintain an acid soil of pH 5. In the South, work with *V. ashei*, the rabbiteye blueberry.

Cherry (*Prunus* spp.). Propagation is usually by budding on rootstocks of the mazzard cherry (*Prunus avium*)—a sweet cherry escaped from cultivation—or on the mahaleb cherry (*P. mahaleb*), with mahaleb 'St. Lucie' best for irrigated alkaline soils. For cherry tree windbreaks use Hansen bush cherry (*P. tomentosa*), which is also good for dwarfing rootstock. To insure a crop, grow several; a sour, perhaps 'Early Richmond' or 'Montmorency', and fairly good pollinators like 'Black Tatarian', 'Grant', 'Lyons', 'Stella', or 'Seneca'. Sour cherries (*P. cerasus*) are more self-fertile than sweet

(*P. avium*). Unfortunately the popular 'Bing' and 'Lambert' are no help for pollination. To work with natives, try sandcherry (*P. besseyi 'Baileyi'*), wild black (*P. serotina*), wild red (*P. pensylvanica*), or chokecherry (*P. virginiana*).

Because many cherries are incompatible, emasculation is not needed before cross-pollinating. If compatible trees are in an area which bees frequent, however, use cages or emasculate. Seeds in cherries mature before fruit, so at harvest, pick and remove seeds at once. Wash and scrub right away.

Sometimes early-ripening seeds germinate inside fruits, because one layer has dried out and then become wet again and started the characteristic water uptake that triggers germination. Late-ripening cherries need cold stratification 3 or 4 months, with stones cracked before or afterwards.

Plant shallowly as soon as germination begins. Set out after 2 months, or put indoors or in a greenhouse. Sweet cherries fruit the fourth year; sour, the third.

In breeding, aim for hardiness of sweet cherries, productivity, increased fruit size, drought-resistance, and disease-resistance. When experimenting with crosses, remember that small cherries are dominant over large, heart-shaped fruit over oval or oblate, and probably sour over sweet. You can get improved stock of *P. avium* and *P. mahaleb* to help you along, and if you are interested in making your own dwarf cherries, get some 'Northstar' rootstock. Professionals use 'Montmorency' as interstock over *P. mahaleb* to speed up bearing. They are also working now to find cherries with a superior ability to take up nutrients.

Chestnut (*Castanea* spp.). The big challenge to the propagator is to discover a new American chestnut that is resistant to the blight which has just about eliminated this species. People now use the highly resistant Chinese chestnut (*C. mollissima*), first introduced about 75 years ago. New varieties available include 'Meiling', 'Nanking', 'Kuling', 'Crane', and 'Abundance'. The Japanese chestnut (*C. crenata*) can be used, too.

Since chestnuts are self-incompatible, breeding is difficult. Although the catkins are insect-attracting, the female flowers have long styles like wind-pollinated species, and chestnuts do pollinate by wind. The American chestnut (*C. dentata*) has been crossed with *C. crenata* to produce fairly blight-resistant 'Sleeping Giant' and 'Essate-Jap'. Also available for breeding is a small Chinese chestnut with very fine nuts, 'Kelsey'. (These three come from the Connecticut Experiment Station in New Haven.)

Budding, grafting, and growing from seed are all possible. The seed needs stratification for 3 months at 40°F. Then sow in light soil with a pH of about 5.5, planting the seeds on their sides, and protecting them from rodents, especially for fall sowings. One good method is to use a #2 tin can, with one end removed and the other slit by two cross-cuts and the flaps turned back. Push down a can with the cross-slit uppermost over each seed planted to keep out rodents and squirrels.

Spring sowings can be made if nuts are stored over the winter in a cool humid room until the burs open, and then afterwards at 35°F. in nearly dry peat moss in plastic bags. With the right conditions, they stay viable 3 years. Seedlings grow to bearing size in 7 years or so, though grafted trees can bear the second year after grafting. Some mature trees bear early, some mid-season, some late. A good early-bearing Chinese chestnut to use for crossbreeding is 'Meiling'.

Cranberry (*Vaccinium oxyoccus*). These slender, creeping, evergreen vines, which grow in bogs, have some good, fairly old clones for propagators to know about: 'Early Black', 'Howes', 'McFarlin', and 'Searles'. Some newer clones are: 'Beckwith', 'Bergman', 'Franklin', 'Wilcox', 'Stevens', and 'Mammoth'.

For stem cuttings take stems with flower buds, in

November. These can be stored a few days in re-frigerator, then rooted in a mist chamber, and grown in nutrient culture of seaweed emulsion solution after rooting. Or cuttings can be pressed into the surface of the cranberry bog. If grown on to maturity, seeds can be collected after hand-pollination.

To pollinate, either take pollen and pollinate stigma at once, or store a short time at 45°F. and emasculate flowers at full bud stage. Stigmas receptive as soon as flowers open. Plant blooms for 4 weeks. When polli-nated, cover with cheesecloth bag. (Pollen tube takes 24 to 72 hours to reach egg and fertilize it. Fruit matures 3 months after pollination. Yield may be 10 to 50 seeds per flower.)

Harvest ripe berries, squeeze out seeds or use blen-der. Or store 6 to 12 weeks at 45°F. and then extract seeds and sow at once. Warmth and light both helpful for germination, usually in 11 to 24 days. Transplant to pots at 3 weeks, fertilize during winter, and give extra light. Harden off before putting out, for 1 or 2 months at 42° to 48°F. Bigger, better plants result from cross-pollination than from self-pollination.

Currants and Gooseberries (*Ribes* spp.). The European black currant (*Ribes nigrum*) should not be grown or propagated in areas where white pine grows because it is the alternate host of white pine blister rust; and the red winter currant (*R. sanguineum*) should not be grown or propagated where wheat is produced because it is the alternate host of black stem rust of wheat. The same is unfortunately true of clove currant (*R. odoratum*).

If you live in other areas of the United States (or in England), you can breed currants from seeds quite easily. Cultivars recommended include 'Baldwin', 'Black Naples', 'Black Grape', and 'Boskoop Giant'. Recom-mended red currants include 'Ruby Castle', 'Houghton Castle', 'Victoria', 'Fay's Prolific', 'Red Cross', 'Red Lake', and one that is supposed to be resistant to leaf spot: 'Rondam'. Because of white pine blister rust, we are not permitted to grow either currants or gooseberries in the region where I live, but if gooseberries were allowed, I would try 'Remarka', 'Reverta', and 'Robustenta', all considered good to start with. Cultivars considered almost or quite spineless are the 'Finik', 'Samburskij', 'Afrikanec', and 'Novosibirskij Valikan'.

Breeding objectives for all gooseberries include spinelessness and resistance to leaf spot and mildew. For currants the aims are high vitamin C content, ease in picking, and hardiness, especially hardiness of the flowers in early spring.

Flowers are hermaphrodite, with male and female parts in one flower: in some the stigma is below the anthers and in others the stigma juts out beyond. The currant 'Baldwin' can self-pollinate, but others need insects. Black currant pollen is sticky; there is no wind pollination. Emasculate both black currants and gooseberries in the bud stage. Pollen is applied to the stigma within the next day or so, usually with a brush. Put bags on those needing protection from visiting insects. If the pollen must be stored, keep it dry and do not attempt to keep it more than 10 to 12 days, unless you use calcium chloride to keep it dry.

When fruit is ripe, harvest, blend, and separate seeds from the pulp. (Gooseberry seeds are sometimes removed from the fruit without blending.) If sown at once, black currant seeds will germinate in 7 days. Otherwise, alternating warm and cold stratification is needed. But red currant and gooseberry seeds must be stratified at 40°F. Give extra light and heat after planting.

Pruning and rejuvenation especially successful on 'Coronet', 'Consort', and 'Brödtorp', which send up good new shoots for cuttings. Softwood cuttings of black currants need to root under mist. Others succeed from hardwood cuttings. Best for cuttings: 'Wellington xxx' and 'Seabrook's Black' (black currants); and 'Bedford Red' for a gooseberry (even if only about half of the cuttings root, for that is really good for gooseberries).

For breeding currants for frost-resistance, use 'Consort', 'Roodnop', and 'Amos Black' to cross with Scandinavian and Russian varieties. Late flowerers are 'Consort', 'Coronet', 'Crusader', and 'Boskoop Giant'. Vitamin C is tops in 'Laxton's Giant', 'Baldwin', 'Raven', 'Boskoop Giant', and 'Laxton's Tinker'. Currants and gooseberries can also be propagated by dividing the plants and by the convenient method of mound layering. (See chapter 3.)

Fig (*Ficus carica*). One of the strange multiple fruits, with "fruits" which are really receptacles or stalks, and what we call the seeds the real fruits. Some varieties require no pollination for the receptacle to swell; in another, the Smyrna fig, a small wasp has to help matters along. She squeezes in a hole in the tip of the fig, passes over the pollen-bearing staminate flowers inside the pouch and on to the styles of the female flowers, where she lays her eggs. Then the nutlet which we call the seed develops. The mature wasps emerge just at pollen time, so they exit with pollen on them, ready to pollinate the next fig flowers they visit. The process is called *caprification*. Flowers of edible figs produce no pollen, so the wasp is a necessity. Caprifigs have three crops a year. Those now in use include 'Samson', 'Stanford', 'Brawlye', 'Roeding', and 'Milco'. Edible cultivars include 'Smyrna', 'San Pedro', and 'Calimyrna', and the early fresh-fruit market figs for California: 'Dauphine', 'Drap d'Or', 'Pied de Boeuf', and others. These may have one or two crops per year. 'Kadota', 'Mission', 'Beall', and others have come down through the generations by rooted cuttings.

Aims in breeding are a good quality fruit that does not require caprification; and one which has white flesh, a golden skin, amber pulp, and the good flavor of the 'Calimyrna'. There is also a need for a fig with such tightly closed pores that no infecting organism can get in.

Dealing with pollination is very complicated, and I must refer you to specialists and their books and publications for explanations of procedures.

Sow fig seeds soon after collection in August or September. Set out the little plants in January (in moderate climates) for their winter dormancy.

On their own roots, figs take 5 to 7 years to mature, but for earlier maturity, graft good spring buds or shoots from 5- to 25-year-old trees onto the juvenile rootstock. If the seeds germinate well, you can get 20 seedlings for budding. Be prepared: half your figs will not be fruit-producing but caprifigs. Some figs will mature, however, without pollination by the wasp. All you have to do is blow into them. Breeding aims include reducing all these complications and developing good species that come to fruit early, mid-season, and late. You may even want to breed seedless (or fruitless) types. There are many opportunities for experimentation.

Filbert (*Corylus* spp.). The European filbert or hazel (*C. avellana*) has been naturally hybridized so often it has many genetic characteristics in the different varieties. Cultivars 'Rush' and 'Winkler' of our native *C. americana* have been used in many crosses in this country.

Propagation by suckers is very easy; after they root, all you have to do is remove them. Softwood cuttings also possible.

Grafting and budding are used in developing new varieties. Rootstocks favored include the ornamental Turkish tree hazel (*C. colurna*), which does not sucker; it is hardy in the North and in Canada, and has a small, fleshy-husked nut. The two American species (*C. americana* and *C. cornuta*) are both disease-resistant. Layering can be used with 'Barcelona' and pollen from that tree has been used with 'Nonpareil', a clone of superior quality.

Growing filberts from seed presents certain problems. The trees are never dormant and bloom from November to March, with the peak in January and February. But there are three rhythms. With some the males bloom first; with others the females; and with some both at once. To breed filberts for cool climates you must use late-blooming parentage. The male catkins of 'Barcelona', though they begin to be visible in June, do not mature to shed pollen until January or February, rather a long growth period that's quite a drain on the tree. Shedding comes on a day when there is a drop in humidity. The female, pistillate flowers also begin to develop in June and mature by August. (Then the tree rests until January or February, though nuts have been known to set from December to March.)

Cross-pollination is necessary, and for 'Barcelona' flowers the favored pollen parent is now 'Daviana'. To prevent wind pollination, put the filbert tree in a tent of plastic on a wooden frame, preferably whitewashed, for it must remain on the tree from December to April (in the main growing area, Oregon). Take the catkins out, and go into the tent to pollinate several times in case you miss some flowers the first time. To collect pollen, it is a very simple matter to bring in laden branches, put them in a jar of water standing on paper and leave them there overnight. The next morning the paper will probably be covered with pollen. Pollen can be stored at 0°F. with no great loss from freezing, but cold 10 to 20 Fahrenheit degrees below that will kill most species of pollen grains. The pollen germinates best at about 60°F., but at 75°F. it will fail.

If you pollinate in January or February, there is tremendous growth by June and nutfall comes between September and October 15. By then the tip of the nut is a year older than its base, for the female flower has been growing for 1½ years—very unusual.

When mature, filbert seeds must be stratified at 40°F. for 3 months. (Follow the directions given in chapter 6.) Seeds that do not germinate the first year may do so in later years. Plant the seeds in deep flats, tall pots, or raised beds.

Gooseberry—see currants

Grapes (*Vitis* spp.). Propagation of grapes is a fairly complicated process, and you're advised to get special help. But in brief, pollinating and growing grapes from seeds involves collecting and tapping, brushing, or blowing pollen grains onto the grape blossoms' receptive stigmas sometime between 6 to 9 A.M. or between 2 to 4 P.M. when the temperature is rising. All this is done at time of stigma receptivity unless pollen previously was stored at 28 percent humidity at 54°F. Germination is often poor, but if fruits mature, extract seeds (as soon as ripe or a little later) by blender, which helps to scarify seeds a little. Plant fresh or stratify at 40°F. for 4 to 5 months. The natural indoleacetic acid in the seeds is thought to multiply if the seeds are stratified. Time to fruiting usually 3 years. Buds from seedlings can be grafted onto bearing vines.

The first breeding aim is the production of clones which will not be sterile. Another is to breed for hermaphroditism. (Grapes are now male, female, or hermaphroditic.) Other aims are for vigor, hardiness, resistance to the disease called *phylloxera,* and tolerance for lime.

Traits of cultivars include: keeping quality—'Flame Tokay', 'Emperor', and 'Malaga'; pleasant, faint flavor—'Almeria' and 'Alphonse LaVallee'; seedlessness—'Thompson seedless', 'Perlette', 'Delight', 'Beauty Seedless', 'Emerald Seedless', and 'Ruby Seedless'. For raisin grapes try: 'Sultanina', 'Black Corinth', and 'Muscat of Alexandria'. Wine grapes include: 'White Riesling', 'Chardonnay', 'Pinot Noir', and 'Cabernet Sauvignon'. Those with good acid-sugar ratio: 'Rubired' and 'Royalty'. For juice try: 'Concord' (not suitable in warmer regions).

Lemon (*Citrus limonia*). Methods of propagation are about the same as for oranges, but in grafting, sour orange rootstock is not recommended: the scions tend to overgrow and bulge out, causing a weak graft union. Use trifoliate orange, sweet orange, or citrange instead.

Lemon scions may also be topworked onto other citrus species if you want to make a grove on one tree. Cultivars preferred in the cool districts along the coasts are 'Lisbon' and 'Eureka'. A dwarf lemon is 'Meyer' and two lemons with very mild acidity are 'Millsweet' and 'Dorshapo'. In the Southeast, where fungus diseases are a problem, try 'Meyer' and 'Villafranca'. Lemons have the pleasant habit of blooming all year.

Lime (*Citrus aurantifolia*). These sensitive trees are either Mexican or Tahitan (Persian) in origin. Mexican lime— also called true lime—is propagated as oranges are, especially by seed. But 'Beares' and 'Rangpur' limes are propagated vegetatively because the first is seedless and the second will not come true from seed as the Mexican will. 'Rangpur' is considered a good rootstock, both for limes and other citrus fruits, but it is too cold-sensitive for some areas. Sparse as a bearer, with only a single terminal bloom, and but a small number of fruit.

Oranges and Some Other Citrus Fruits. From seed it is very simple to grow the sweet oranges (*Citrus sinensis*), the sour orange (*C. aurantium*), also mandarins (*C. nobilis deliciosa*). Luckily most hybrids will breed true because the seedlings develop, not from an egg, but from a tissue in the embryo sac near the egg. In this process called apomixis, no union of sperm and egg takes place. And more than one embryo will develop from a single seed. Put a seed on a wet blotter and watch it develop. Then plant (without drying) in a flat of light soil. It will be a thorny tree for that is the original, unhybridized state of orange trees. Hardiest citrus fruits are trifoliate ones or a cross of those with sweet oranges called citranges, with cultivars 'Morton', 'Rusk', 'Savage' (or for rootstock, 'Troyer'). All of them hardy up into the Carolinas. So are kumquats, and their cross with trifoliate oranges. Germination is slow, probably 2 to 4 weeks at 65° to 78°F. but you can hasten it by removing the seed coat or slitting it. Use sterile knife or sterilize your fingers with Clorox® solution. You can grow the seedlings, bud graft them to royal lemon (*C. limon*) or to their own strong seedlings as stock.

Pollination is, however, needed. Citrus plants flower in two ways: for controlled pollination, remove the undeveloped buds when you pollinate because the anthers are near the stigma, and self-pollination is frequent. Most pollination, however, is by bees attracted by the scent, color, and great supply of nectar. To collect pollen, choose flowers on branches bearing moderately, not heavily, using the larger, earlier flowers with full-sized pistils. Pick the whole flower just before opening and remove pistil and petals. Use what's left to pollinate emasculated flowers, or store the pollen, which is shed within a day after picking. But storage is rarely necessary for blooming period is long. Keep the pollen dry and cool until used. Fruit for seed is harvested when mature. When you harvest mature fruit, cut only through the rind and then twist the fruit open. Pick plump seeds, wash well, and to get rid of possible brown rot fungi, agitate seeds for 10 minutes in hot water at 125°F. Wipe dry. But do not let seeds dry out, even for a few hours. And do not soak. Store or grow under the most sterile conditions you can manage.

Grafting seedlings will present some problems. Since it takes 5 years for seedlings to produce fruit, they are in a juvenile state for a long time, so buds taken as scions will retain the juvenile characteristics of thorniness, vigorous upright growth, and slowness to fruit, especially if taken from the younger, lower part of the tree. Other possible defects include hollow cores, thick puffy rinds, and unattractively long fruit. So let seedlings grow five years before budding and grafting. Older trees can be topworked. Cuttings will succeed if mist-sprayed and kept moist.

Cultivars to try include 'Washington Navel' for the Southwest; 'Temple', 'Hamlin', 'Jaffa', 'Orlando', and 'Parson Brown' in the South and Southeast; or 'Valencia' in most climates where oranges grow.

For tangerines, try 'Clementine' and 'Dancy'; for mandarins, 'Owari Satsuma' and for tangors, try 'King'. Rootstocks from the sour oranges 'Oklawaha' and 'Bittersweet' have proved useful though they cannot stand temperatures below 20°F.

Papaw (*Asimina triloba*). This tree is native to southern Ontario and the eastern United States, but not New England. *A. speciosa,* of southeastern Georgia and northeastern Florida, is said to yield a small, delicious fruit. To grow papaws from seed, cross-pollinate and pick fruit just before ripe. Put seeds in cold stratification at 40°F. about 3 months. Keep wet. Germination will be difficult; seeds planted in the fall in rotted sawdust and kept over the winter in a cold frame may do better. Do not try to transplant seedlings unless plants are very small.

Grafting fairly successful, using whip graft. Suckers come, but are not easy to root. Root cuttings taken at end of frost in spring may be more successful. Cultivars to try include 'Davis', 'Overleese', 'Fairchild', and 'Ketter'. The first two have good flavor; the second two are early ripening and also good flavored. Breeding aims can include overcoming slowness of germination and early growth; development of self-fruitful cultivars; and further study of papaw's mycorrhizal association needs.

Peach (*Prunus persica*). Experts seem to think that just about the best of all fruit-tree breeding has been the work done with peaches. Disease problems have been relieved; the genetic causes of various traits such as climatic adaptation, productivity, and winter hardiness are well understood; and by now the future directions are clear. But propagating is still an inviting field for amateurs and professionals alike.

These trees grow well in the South where there are at least 500 to 1,000 hours of winter cold below 40°F. to break bud dormancy; and in the North near big bodies of water (to modify the climate). For hand-pollination, seed care, and planting details see chapter 6, page 124.

Remember that with peaches the pollen and ovules develop at time of bud-swell, but the stigma is receptive at time of bloom. Collect pollen at balloon stage when flowers still turgid. Dry, and use that season or save a year in deep freeze. Remove male stamens from flowers you plan to pollinate, but let no pollen shed into flower while working. Pollinate when stigma has fluid on it and temperature is above 60°F.

When fruit is ripe, remove flesh. Stratify seeds by mixing in moist peat or peat-sand mixture and placing at 40°F. for 3 to 4 months. (Removing seed from stone enhances germination.)

Plant seeds in sterilized sand in warm, moist place. Move once to nursery, and then to orchard at 5 to 7 leaf stage. See that soil has enough lime. Shade seedlings to avoid wilting. These stock seedlings must suffer no setbacks.

Graft in June in South; in August in northern states. For scions use cuttings or buds from compatible clones. For example, use scions from progeny of Chinese cling peaches such as 'Elberta', 'Belle of Georgia', 'Hiley', and 'J.H. Hale' crossed with such other *Prunus* species as *P. davidiana, P. ferghanensis, P. kansuensis, P. mira,* and the Canadian clones 'G.F. 677' and 'G.F. 557'. For dwarfing (plus vigor) try 'Harrow Blood', which dwarfs by 20 percent, and 'Siberian C', by about 15 percent. Both these are also good for early bearing, early foliation, and bud hardiness (valuable characteristics when breeding peaches for cold climates).

Plum seedlings can be good for vigor also. Use for rootstocks such clones as the 'Wayland' plum (*P. hortulana baileyi*), which is good for dwarfing; and the F_1 hybrids with minimum cold weather requirements of 600 to 650 hours of cold. Try crossing one of those with the tender 'Earligold', 'Maygold', or 'Springtime'. Hardier are 'Elberta', 'Golden Jubilee', 'Redhaven', and 'Sunhigh'. Among the hardiest are 'Veteran' and 'Prairie Dawn'.

Softwood cuttings very easy.

(To help your peaches survive, water well in the fall, use a cover crop of rye or vetch, and pick off excess fruit after the June drop. Do not overfertilize.)

Pear (*Pyrus communis*). Breeding of pears in this country now is so focused on hardiness and resistance to "pear decline" and viruses that European pear-breeders are way ahead on quality. Gives chance for amateur breeders to work for quality in United States, too. In Europe, the dwarfing understock quince (*Cydonia oblonga*) clone used is Malling Quince A; but in U.S., 'Quince B' and 'Quince C' are best for dwarfing. Heat treatment frees them of virus. Many pears self-unfruitful and incompatible as are 'Bartlett' and 'Seckel'; but with the blight-resistant 'Old Home' as interstock in double-grafting, both are compatible with quince understock, so use virus-free quince with 'Bartlett', 'Seckel', 'Bosc', 'Clapp's Favorite'; and others. You can also use quince seedlings or seedling stocks of 'Bartlett', 'Bosc', and 'Winter Nelis'; and for scions: 'Bartlett' and 'Bosc', plus 'Anjou', 'Comice', and 'Passe Crassane'. For hardiness breed with *Pyrus ussuriensis*. This and the ornamental 'Bradford' callery pear (*P. calleryana*) are both very valuable as they are resistant to the devastating fire-blight.

Collect pollen by cutting and forcing branches in dormant or early tight-bud stage 2 or 3 weeks before outdoor bloom. Harvest just before pollen to be shed, and proceed as described in chapter 6. Dry at about 75°F. and store very dry at 40°F. until needed. Emasculate the basal pear blossoms. Do not use big king blossom or small terminal ones.

Harvest fruit for seeds when not yet ripe. For wet stratification sow outdoors in seedbed or keep moist in indoor cold storage at 33° to 34°F. (for 60 days in warmer climates and 90 days in colder). Pasteurize the storage medium. Germination in 7 to 10 days at 68°F.

Then begin to add nourishment and when plants reach 5 to 6 inches, transfer to richer mixture with pasteurized soil.

Pears stay juvenile a long while; to induce earlier maturity, graft to older stock, and grow rootstock seedlings as fast as possible with extra light, carbon dioxide, and fertilizer. Give plenty of room and do not prune.

Dwarfing as breeding aim is helped by cultivar 'US 309', which has a natural dwarf habit. Fertilization must come from compatible cultivars. Good dwarfs to try for eating quality include 'Angouleme', 'Diel', 'Easter', 'Glou Morceau', 'Louise Bonne', and 'Elizabeth'. Varieties needing double-working if attempted as dwarfs include 'Bosc', 'Winter Nelis', 'Washington', 'Sheldon', and 'Dix'. Those best left as standards and not dwarfed include 'Bartlett', 'Seckel', 'Gray Doyenne', 'White Doyenne', and 'Summer Doyenne'; but some good for either are 'Anjou', 'Boussoc', 'Tyson', 'Colmar', and 'Superfin'. Consult your county agent for information about hardiness and suitability for your area.

Pecan (*Carya illinoensis*). Pecans are good nut trees for the South, with recommended varieties: 'Schley', 'Stuart', and 'Success'. Texas-bred varieties include: 'Barton', and the series named for Indians: 'Wichita', the favorite, and 'Cherokee', 'Chickasaw', 'Cheyenne', and 'Sioux'. Shade-tree pecans for the North include 'Burlington', 'Desmoines', 'Pleas', 'Green River', and 'Starking Hardy Giant'. Grow two varieties near each other. Propagate by softwood cuttings under mist, grafting or budding, or seeds.

The catkins bearing pollen grow from buds of the previous year's growth. The female, pistillate flowers are nearby. Pollen-shed lasts 5 to 50 days; sticky stigma receptive 6 to 14 days, with pollination peaking on hot, dry afternoons; but it may take 2 to 6 weeks before fertilization, and fertilized egg may not begin to divide for 8 weeks. To interrupt natural pollination, use a bag of sausage casing to cover the flowers for a while before receptivity of the stigma, as the professional breeders do. Pollen, collected from catkins, is used as soon as dry. But can be stored at 15°F. and 25 percent humidity for a year. Sausage case is plugged sometimes with wad of cotton, and pollination done by pushing hypodermic needle through sausage casing and blowing pollen in from pollen-filled bulb. Time for this when stigma is brown and glossy. Beware of too much pollen; causes nut drop.

Germination can begin while nuts still on tree, because the natural growth-inhibitor peaks about 20 days before harvesttime. Plant (in South) any time between harvest and the following spring. In North stratify at 32° to 34°F., in slightly acid peat moss; 10 weeks time is shortened if shells removed, but never let nutmeats dry out. Pecans do not always come true from seed.

To graft, use large or fairly large trees for stock and 5½-month-old buds of seedling branches (with reddish leaves, erect growth, and no flowers). Use T-bud, H-bud, or whip graft of whole seedling. Wait 8 years for results.

In breeding, work for lessened nut drop, hardiness, and for disease-resistance by including wild nut stock; also for precocity and productivity. Some existing early-bearers now produce excessive crops of small nuts. Try to work with others such as 'Desirable', 'Elliott', 'Farley', and 'Success'. Other good precocious varieties are 'Cherokee', 'Shoshoni', 'Brooks', 'Candy', 'Evers', and 'Western'. Sturdier late-bearers to crossbreed with include 'Schley', 'Mahan', and 'Curtis'. (These are not stolen much by squirrels.) 'Candy' and 'Mahan' are susceptible to scab, but 'Gloria', 'Elliot', and 'Farley' are not.

Pecans belong to the hickory group of the walnut family. Crosses with shellbark hickory, called micans, make very good, vigorous shade trees that sometimes produce nuts. Varieties considered good are 'Gerordi', 'McCallister', and 'Koon', and the more fruitful 'Burlington' and

'Bixby'. In making such crosses for improved compatibility, the best shellbark hickory to use is 'Weiper'; and the shagbarks are 'Hales', 'Kirkland', and 'Kentucky'. (So far pecans grafted onto hickories have not been very successful.)

Hopes are to get pecan trees hardy up to Zone 4 but not subject to nut drop. Varieties to include for these aims include: 'Green River', 'Butterick', 'Indiana', 'Busseron', and 'Niblack'. These are all very good lawn trees. Cultivars favored for rootstock for all aims are 'Riverside', 'Mahan', 'Apache', 'Hollis', 'Moore', and 'Western'. (Remember that pecans have long taproots and are very hard to transplant.) Cuttings and air layering rarely successful.

Wild nuts in the hickory group already tried for crossbreeding include: pignut (*Carya glabra*), fairly hardy; shellbark hickory (*C. laciniosa*), not quite so hardy, but found from New York south to Tennessee; shagbark hickory (*C. ovata*), and mockernut (*C. tomentosa*), both hardy. Some have very thick shells and very small meats, so crossbreeds with pecans for good traits may be badly marred by the appearance of these poor characteristics.

Persimmon (*Diospyros virginiana*). This American persimmon, plus a garden cultivar from China, *D. kaki,* are propagated easily. Adapted to climates from Florida to Canada, *D. virginiana* is usually grafted, but root cuttings are easy to grow. Seeds also. These plants have pistillate and staminate flowers, and they are pollinated by wind or bees. Fruits ripen in August in the South; in fall and until December in the North. Stratify seeds at 40°F. for 3 months before planting. Trees start to bear at 3 or 4 years. Grafting on seedlings with T-bud graft, spring, or side-bark grafts on stock about 1 inch in diameter. Use water sprouts or top twigs. Old trees topworked will bear in 2 to 3 years. Persimmons have taproots; very difficult to transplant.

Plum (*Prunus* spp.). Many species available to be used in breeding and propagation. *Prunus americana* a hardy one; *P. angustifolia,* disease-resistant; and *P. cerasifera,* the cherry plum, can all be used for rootstocks. A Canadian plum (*P. nigra*) is also hardy; and *P. domestica,* very productive, is the source of many good plum cultivars. Better for the South are 'Chickasaw' and *P. hortulana* 'Bailey'. And there is the Damsun plum (*P. instititia*), cultivated for 2,000 years. Luther Burbank, the best-known plum breeder, used *P. salicina,* a Japanese plum, and among his cultivars was the now common, excellent commercial plum 'Santa Rosa'. Present breeding aims are resistance to disease and pests, and extension of climate adaptation, with spread of growing season from April to October. Parental material available for solving many problems.

Collect pollen for hand-pollination just before the petals open and emasculate unless using a self-unfruitful cultivar. Pollen can be stored at 25 percent humidity at 35°F. for a year (or more). Otherwise apply as soon as stigma ready. Keep applying every few days throughout the blooming season. Cover unemasculated flowers with paper bags or cheesecloth to prevent bee activity.

When ripe, harvest and remove seeds. (Early-ripening varieties most difficult to grow; may not germinate.) Crack harvested seeds right away, and stratify for 2 to 3 months at 40°F. in acid peat or sphagnum moss, rinsed with weak Clorox® solution and twice with clear water. If once dried, later water uptake likely to lead to infection.

When signs of germination appear, remove and plant seeds. Keep cool over the winter, or harden outdoors to force bud dormancy, then put in garden in spring. Seedlings can be used for grafting or grown on. Propagate the good ones by budding and grafting.

Remember that self-unfruitfulness is common in plums, though 'Santa Rosa' and Damsun plums are partly self-fruitful. Japanese-American crosses, such as

'Compass', 'Kaga', and 'South Dakota' are fairly self-fertile. The very mixed parentage means many surprises for the breeder, and many pleasures.

For good flavor and quality, try *P. domestica* as a parent. For disease-resistance, try crosses of American species such as 'Bruce', 'Munson', 'Opata', 'Burbank', 'Kelsey', or 'Santa Rosa', and 'Green Gage', or 'Simon'. The 'Green Gage' (or 'Reine Claude') is reliably self-fertile; and so are 'Stanley', 'California Blue', and 'Yellow Egg', as well as two Damsuns: 'Shropshire' and 'French'. These Damsuns along with the cherry plum and the Japanese *P. silicina* are the most hardy.

Use the cherry plum or myrobalan (*P. cerasifera*), or a cross of this with *P. munsoniana*, the marianna plum, for rootstocks. For European plums use *P. cerasifera*; for Japanese use *P. persica* and for American use *P. americana*. For dwarfing, try *P. besseyi*, the native American sand cherry. There is a wide range of possibilities now open to the experimenter.

Pomegranate (*Pumica granatum*). Inferior seedlings easy to grow from seeds pressed out of pulp of fruit and planted in flat and lightly covered. Propagation by hardwood cuttings better, though layering and grafting are also used. If you graft, use pomegranate seedling rootstocks. Prune out water sprouts and suckers.

Quince (*Cydonia oblonga*). Mound layering. Hardwood cuttings with heel, from 2- to 3-year wood, which has burs on it with many adventitious root initials inside. If you have to use 1-year wood, use bottom part, not terminal. Self-fruitful, but seedlings sometimes difficult. Budding on quince layers (or seedlings). Good cultivar 'Orange', but also try 'Champion', 'Pineapple', and 'Smyrna'.

Raspberry (*Rubus* spp.). Cultivated forms include the European *R. idaeus*, and the American *R. strigosus*. The blackcap or black raspberry is *R. occidentalis*, a native plant. Black raspberries tip layer by themselves, but should be prevented from layering all shoots. Red raspberries propagate by underground roots, with shoots. Mark and save the best, leaving some of old root attached for spring or fall transplanting. (Cut back if too tall.) Suckering shoot growth can be stimulated by cutting the roots and also by mulching. Root cuttings easy; make them 3 inches long.

As with some other *Rubus* species, fruit normally comes on biennial canes (with dormant buds followed by fruit buds, which develop the following spring, except for fall-fruiting types such as 'Heritage'). All old canes should be removed after harvest. Keep only sturdy new ones.

Flowers are of four types: male, female, hermaphroditic, and neuter. Buds form on first-year canes, and in the spring the male parts form after the others, though female flowers may develop a month after male flowers. Therefore remove and store pollen up to 4 weeks at 40°F. Keep pollen dry; if necessary, with desiccator such as calcium chloride. Keep all utensils sterile. When stigmas receptive, pollinate three times a day. Protect with white paper bag. When berries ripe, harvest and use blender with plenty of water to separate good seeds, which will drop to the bottom. Air dry and store at 32° to 40°F.

Seeds hard to germinate because of three causes of dormancy: impermeability to air and water; hardness of seed coat, which resists swelling of embryo; need for cold stratification for embryo to develop. Therefore scarify seeds by shaking with coarse sharp sand, plant in moist vermiculite or shredded sphagnum moss, cover very lightly, and store moist for 3 to 4 months at 35° to 40°F. Then move to 70°F. Do not move until fibrous roots appear.

Breed raspberries for good vigor, strong and sturdy first-year canes, erect growth, and good flavor. (Commercial breeders also aim for attractive bright-colored seeds that show up prettily in jam.) In North, experiment with cultivars 'Cuthbert' and 'Latham' if your soil is light or 'Sumner' and 'Malling exploit' if soil is heavy. In South, try 'Southland' and 'Dormanred'. Black raspberry parentage might include 'Bristol' and 'Cumberland'.

Strawberry (*Fragaria* spp.). Aside from the familiar wild strawberry (*Fragaria virginiana*) and the cultivated strawberry, including all the progeny of *F. virginiana* x *F. chiloensis*, there are the Himalayan (*F. daltonia*) and the Alpine strawberry (*F. vesca*) as well as *F. viridis*, *F. orientalis*, and various crossbreeds. Everbearing strawberries are twice-bearing, really, once in the spring and again in late summer and fall.

Everbearing varieties are convenient to propagate, for all runners should be removed from the mother plant anyway, not only during the first year (when the blossoms are removed, too) but throughout the life of the plant. Use as cuttings. One-crop strawberries increased by placing flowerpots or cans in the soil during June and July under whatever runners you see, and letting them root. Then cut them loose, move in the pot, and replant in a new bed, with the roots well spread out as in any strawberry planting. Keep crown above the level of the soil surface. One-crop varieties, typical of the North, have many runners. Everbearing, more typical of the South, have fewer.

Though wild strawberries may be female, male, or hermaphroditic with both sexes in one flower, modern cultivars are hermaphroditic. For breeding by seed, the flowers will need to be emasculated preceding hand-pollination. Remove anthers, petals, and sepals, and be very careful to spill no pollen on any of the many pistils, each of which develops a "fruit" or what we call a "seed." Bag the emasculated flowers. Pollen may be stored for several days at 50°F. if necessary. About a month later, ripe fruits can be harvested and the seeds removed at once. Blend in blender 12 seconds with water. Let stand until seeds fall and pulp floats to the top. If storage necessary, the seeds will remain viable for 20 years. Plant in flats. Germination may take several months, though a few seeds will come up in a week or so. Better germination will come from cold treatment for 2 to 3 months, and by exposure of seeds to light. (Scatter on sphagnum moss on pasteurized sandy soil.) Transplant after 6 weeks, harden, and set out in field after 6 weeks more. When you do get a good cultivar, always propagate it thereafter by runners.

F. ovalis is valuable for hardiness, and *F. virginiana* for blossom hardiness, while F. *chiloensis* is for disease-resistance. Two for winter hardiness are 'Arapahoe' and 'Ogallala', both everbearing. Blossom-hardiness may be found in 'Howard 17' and 'Earlidawn'.

Good yield and good size are two other breeding aims. 'Florida ninety', 'Lassen', and 'Shasta' can be used for these aims in Florida and southern California; and in the northern states try 'Catskill', 'Sparkle', 'Midland', 'Midway', 'Pocahontas', and 'Robinson'. All are self-fruitful, so be careful in pollen-collecting.

Specifically for fruit size, try 'Gorella' and 'Abundance', or crosses of 'Gorella', and either 'Ydun' or 'Senga', as well as 'Belrubi', 'Catskill', 'Guardian', 'Robinson', 'Sequoia', 'Titan', and 'Vesper'. Those lasting well on the vegetable stand have been 'Tioga' and 'Tennessee Shipper', with firmness outstanding in 'NY 844', 'Tennessee', 'Gorella', and 'Pocahontas'. The old-time favorite for firmness was 'Blakemore'. Also are: 'Albritton', 'Apollo', 'Dixieland', 'Earlibelle', 'Holiday', and 'Tennessee Beauty.' Favored for freezing have been 'Fairfax', 'Earlibelle', 'MdUS 2713', and 'Ark 5018', both for color and taste. And for vitamin C the highest are 'Marshall', 'Clarke', and 'Progressive'.

Anyone who decides to experiment with strawberry breeding should look into possible virus and disease difficulties. The literature on these subjects is vast and helpful. Ask your county agent.

Walnut, Black Walnut, and Butternut (*Juglans* spp.). The English or Persian walnut (*Juglans regia*) is propagated by both budding and grafting, usually by whip grafting year-old seedlings or by patch budding of the young trees in the summer, with an understock of California black walnut (*J. hindsii*) or English walnut seedlings immune to crown rot. The black walnut from the Eastern states (*J. nigra*) and the cross (*J. regia* x *J. hindsii*) called 'Paradox' also used. (All other *J. regia* crosses likely to fail at the graft union after 12 to 15 years.) *J. regia* is susceptible to oak root fungus and crown rot. Cuttings are not successful.

But these self-fertile trees can be grown from seed, usually in screened, buried flats to keep out rodents. Walnut catkins produce nearly 2 million pollen grains.

To pollinate, gather catkins when ready, let pollen fall on white paper for 24 hours. Put through a sieve and store if necessary in the freezer: it can last a year. Cover the pistillate flowers while receptive with thick, white cloth bag to keep out windblown pollen. (Clear plastic will cause burning.) Put a plug of cotton inside and use hypodermic needle to blow the pollen into the bag. (Stigmas are ready when the lobes separate but before brown specks appear.)

Seed will be mature *before* the nuts are hullable. Harvest then, and give 6 to 8 weeks of stratification at 35°F. Plant in flats about 1 inch deep and plant out when dormant the first season. When seedlings have been in the nursery for a year, take scions. Graft them onto 2-year-old stocks.

Best cultivars for hardiness are from the Carpathian walnuts. Hardiest of all include 'Colby', 'Broadview', 'Gratiot', 'Hansen', and 'Somers'.

West Coast propagators can work with 'Eureka'; the Santa Barbara soft-shelled varieties 'Placentia' and 'El Monte'; and the thin-shelled 'Farquette', 'Blackmer', and 'Hartley'. A good early cultivar is 'Payne'; a good late one, 'Franquette'.

J. nigra, the Eastern black walnut, which grows nearly down to the Gulf of Mexico, can be propagated by grafting and grown from seed. In some areas it self-seeds readily. Very hard-shelled. The breeding aims include thinner shells, good cracking quality, and productivity. Named varieties to use include 'Thomas', 'Stambaugh', and 'Snyder'.

J. cinerea, the butternut, is a less distinctive tree, with hard and rough-shelled nuts, small kernels, and not so distinctive a flavor as the black walnut. Named varieties include 'Deming', 'Love', and 'Kenworthy', available for breeding experiments.

Ornamental Trees and Shrubs

Abelia, glossy (*Abelia* x *grandiflora*). Sow seeds when ripe; or store in cool place for a year. Leafy semi-hardwood cuttings in summer or fall. Hardwood cuttings, fall or spring.

Acacia spp. Seeds of mimosas and others need soaking overnight in water heated to 180°F. Either let water cool, or use a yogurt maker to maintain heat. Or you can scarify seeds. Semi-hardwood cuttings. Have taproot, so do not transplant, Use peat pots. Good for bees.

Alder (*Alnus* spp.). Sow when ripe or keep moist and stratify at 40°F. for 3 months. Keep seeds very clean. Grafting on *Alnus* seedlings of weeping or featherleaf varieties will succeed.

Aloe spp. Seeds. Suckers. Softwood cuttings.

American arborvitae—see Evergreens

American sweet gum (*Liquidambar styraciflua*). Collect seeds in fall; keep moist; stratify at 40°F. for 1 to 3 months. Budding. Grafting. Softwood cuttings.

Ash tree (*Fraxinus americana*). Stratify seeds for 1 month at 68° to 86°F., then 2 months at 40°F.; then sow. Or store in airtight container in cool place for 1 year; then sow. Some take 2 years to germinate. *F. excelsior* needs 1 to 3 months stratification at room temperature, then 3 to 5 months at 40°F. Budding. Grafting.

Aucuba (*Aucuba japonica*). Collect seeds from female bush and sow at once. Cuttings from softwood or hardwood.

Azalea spp. Usually listed as *Rhododendron* varieties. Layering for a year or so is best method. Cuttings from young evergreen or semi-evergreen azaleas in mid-summer, before wood has hardened. Use 4-inch pieces, removing all but top leaves. Double pot method recommended. See chapter 2. From seed: sow when ripe or store cool for 1 year. Plant late winter or early spring at 70°F. day and 55°F. night temperatures. Germination in 30 to 48 days. Use acid leaf mold, mellowed sphagnum moss (soaked 24 hours), and lime-free water. Protect seedlings for 2 years.

Barberry (*Berberis* spp.). Sow fall or spring, stratifying seeds you keep over till spring for 6 weeks at 40°F. Keep very clean. Cuttings in summer, but not easy. Division of clumps. Attractive to birds.

Bayberry (*Myrica pensylvanica*). This and other *Myricas* all propagated by seeds. Remove all wax, and keep from drying out. Plant in fall. Very attractive to birds.

Beech (*Fagus* spp.). Plant seeds fall or spring, stratifying seeds kept over for 3 months at 40°F. Keep moist. Grafting on seedlings of *F. sylvatica*, the European beech. Keep root-pruned to stop taproot from growing.

Birch (*Betula* spp.). Plant fall; or spring, if stratified 1 to 3 months during winter at 40°F. Stratification not necessary if germination takes place in light. River birch (*B. nigra*) may be sown when ripe. Weeping forms grafted on European birch. Cuttings difficult.

Bittersweet (*Celastrus* spp.). Vines. Keep male and female vines near each other. Remove seeds from berries, plant in fall or stratify 3 months at 40°F. Softwood cuttings in summer root easily. Hardwood cuttings need root-promoting hormone aid. Attractive to birds.

Bottlebrush tree (*Callistemon lanceolatus*). Seedlings from seeds not very successful. Use leafy semi-hardwood cuttings.

Box (*Buxus sempervirens, B. microphylla*). Cuttings of softwood in spring or summer; or hardwood (the more usual method) in winter or spring. If mist used, take anytime. Division. Seeds very slow.

Broom (*Cytisus* spp.). Especially Scotch broom (*C. scoparius*). This one good as understock for grafting other varieties. Sow seed as soon as ripe, or soak in hot water overnight after nicking seed coat. Softwood and hardwood cuttings. Take in summer and use mist and bottom heat.

Buckeye or horse-chestnut (*Aesculus* spp.). Sow at once or keep seeds moist and sow after 4-month stratification at 40°F. But the shrub, bottlebrush buckeye, needs no stratification. Layering of low-growing species. Grafting and T-budding on *A. hippocastanum* as rootstock.

Butterfly bush (*Buddleia* spp.). Seeds do not always come true, but if sown early fall as soon as ripe, will bloom in a year. Cuttings easy. Use either softwood or hardwood.

Camellia (*Camellia* spp.). Difficult. If aim is green bushes, seeds are satisfactory. For flowers, use layering, cuttings, grafting. Cuttings should be taken in mid-summer. Grafts by cleft or bark graft best. Use dormant stock and cover with soil up to leaves of the scion.

Catalpa tree (*Catalpa* spp.). Sow seeds when ripe, or store a year in airtight container in the refrigerator. Softwood cuttings in summer. Root cuttings very easy. Graft for umbrella tree.

Ceanothus and New Jersey tea (*Ceanothus* spp.). Very hard seeds; put in hot water 4 to 5 times their volume, 180°F. up to 200°F. or, if you are careful, 212°F. Let cool in water for 24 hours. Plant swollen seeds at once. Or stratify at 40°F. for 4 months. Softwood or hardwood cuttings easy. Take in July or August.

Chaste tree (*Vitex negundo* and *V. agnus-castus*). Seeds need stratification for 3 months at 40°F. before sowing. Layering. Softwood and hardwood cuttings. Attractive to bees.

Cherry, flowering (*Prunus sargentii, P. campanulata, P. yedoensis, P. serrulata, P. subhirtella*). All budded onto the mazzard cherry (*P. avium*). *P. sargentii, P. campanulata,* and *P. yedoensis,* come true from seed when cross-pollination prevented. Cuttings from any very difficult.

Cherry laurel (*Prunus laurocerasus*). Seeds stratified for 4 months at 40°F. Softwood cuttings. Budding. Very useful shrub for the South, a broadleaf evergreen.

Clematis vine (*Clematis* spp.). Cuttings under mist-spray.

Coral berry (*Symphoricarpus orbiculatus*). Division. Cuttings. Root suckers. Seeds very difficult. Attractive to birds.

Cotoneaster (*Cotoneaster* spp.). Many spp. require double stratification, usually 3 to 5 months at room temperature, then 3 to 5 months at 40°F. Cuttings difficult. Simple layering possible.

Crab apple, flowering (*Malus floribunda* et al.). Process seeds as soon as ripe, especially those that will come true from seed such as *M. florentina, M. hupehensis,* and *M. toringoides,* the cutleaf one. Or provide cool storage followed by 3 months' stratification at 40°F. and then sow. All others asexually propagated by whip grafting or T-budding. Attractive to some birds.

Daphne (*Daphne mezereum* and hybrids). Graft onto *D. mezereum*. Cuttings. Root cuttings. Many daphnes self-layer, especially *Daphne cneorum*. For layering use previous year's shoots in spring; remove from plant the next spring. Do not move the mother plant.

Deutzia (*Deutzia* spp.). Hardwood cuttings easy. Also softwood cuttings in summer, under glass or polyethylene.

Dogwood, flowering and other (*Cornus florida*, *C. alternifolia*, *C. kousa*, *C. stolonifera*, *C. alba* 'Siberica' and C. mas, the Cornelian cherry, so-called). Of these, the seeds of *C. florida*, the flowering dogwood, need stratification at 40°F. for 4 months. But may succeed if planted in fall and protection given the seeds. Same for *C. kousa* and *C. stolonifera*, but all others need double stratification of 3 to 5 months at room temperature and 3 to 4 months at 40°F. Crown grafting and shield budding on *C. florida*. Cuttings from *C. florida*, taken in spring or early summer from new growth after flowering usually succeed under mist. Hardwood cuttings from *C. alba* 'Siberica', Siberian dogwood.

Elaeagnus, thorny (*Elaeagnus pungens*). Broadleaf evergreen related to Russian-olive. Can be grown from seed processed as soon as ripe, if stratified for 3 months at 40°F. Root cuttings in spring. Softwood and hardwood cuttings.

Elderberry, American (*Sambucus canadensis*). Division. Cuttings easy. Seed variable, but 2 months' warm stratification followed by 3 to 5 months at 40°F. Attractive to birds.

Elm (*Ulmus* spp.). In hopes the new cure for Dutch elm disease will be successful, many people like to propagate elms. Try hybridizing the American elm (*U. americana*), the Chinese elm (*U. parvifolia*), the Siberian elm (*U. pumila*), which is also called Chinese elm in some catalogs, or the English elm (*U. procera*). Sow seed when still moist and as soon as ripe. Do this in spring (except for *U. parvifolia*, which ripens in the fall). Or store at 32° to 40°F.

Eucalyptus (*Eucalyptus* spp.). If you live in California, plant seeds in the spring using those from mature capsules not quite ready to open. Viable 10 to 15 years. Beware damping-off, especially with *E. niphophila* and *E. pauci-*

flora. With these and with blue gum (*E. globulus*), no dormancy needed. Cuttings possible only from red gum tree (*E. camaldulensis*), and careful wounding and misting needed.

Euonymus (*Euonymus* spp.). Seed is erratic, but if kept moist, stratified at 40°F. for 3 to 4 months, and grown in greenhouse, success possible. Hardwood cuttings in early spring and leafy semi-hardwood cuttings usually succeed. Use deciduous varieties for hardwood cuttings; evergreen for semi-hardwood, for example, the popular *E. fortunei* and *E. japonicus*.

Evergreens, Needle-leaved. Cuttings often most successful. See chapter 2.

> **American arborvitae** (*Thuja occidentalis*). Stratify seeds when collected for 2 months at 40°F. with high moisture. Softwood and hardwood cuttings easy in most seasons. Remember that juvenile wood produces trees with juvenile characteristics.
>
> **Fir** (*Abies* spp.). Balsam fir (*A. balsamea*) best to grow in North; Fraser balsam fir (*A. fraseri*) in South. Pistillate cones at top of trees shatter when ripe. Gather seeds and plant as soon as ripe, or store in cool place for a year. Some seeds need cold stratification at 40°F. for 2 weeks to 1 month. Beware of damping-off and of hot sun on seedlings. Graft with terminal shoots. Hardwood cuttings possible, but take in winter and use mist.
>
> **Hemlock** (*Tsuga* spp.). Stratify seeds 3 months at 40°F. and then sow, or store 1 year and then stratify. Graft, using *T. canadensis*, the hardy Canadian hemlock, as understock. There are fine dwarf varieties of the Canadian hemlock.
>
> **Juniper** (*Juniperus* spp.). Seeds may take 2 years to germinate. Stratify at room temperature for 3 to 5

months, then at 40°F. for 3 months. Hardwood cuttings easy of Chinese juniper (*J. chinensis* 'Glauci Hetzi'). Graft with these or rooted cuttings of eastern red cedar (*J. virginiana*). (Beware of rust, especially in apple country.)

Pine (*Pinus* spp.). Stratify all pine seeds for 3 months at 40°F. before planting. Germination times vary greatly, even take up to 2 years. Grafting possible on same or closely related species.

Spruce (*Picea* spp.). Sow when ripe, or after 1-year storage in a warm place. No pretreatment necessary. Root cuttings of dwarf Norway spruce (*P. abies*) and use these or rooted cuttings of Colorado spruce (*P. pungens*) for understock when grafting.

Yew (*Taxus* spp.). Japanese yew (*T. cuspidata*) is hardy. But English yew (*T. baccata*) is not hardy and it passes along its tenderness if used as understock in grafting. Gather the seeds which are covered by a red, fleshy structure, from female plants. Stratify at room temperature for 3 months, then at 40°F. for 4 months. Growth of seedlings very slow. Cuttings of new growth with mallet of old growth. Cold frame and misting help. Or use bottom heat and cool air. Male plants of *T. cuspidata* root more easily than female.

False cypress (*Chamaecyparis* spp.). Collect seeds in fall and dry them in warm room (up to 90°F.). Stratify at 40°F. for 2 to 3 months. Cuttings fairly easy, especially of juvenile material, taken in fall or winter. Use side shoots of current year's growth.

Fir—see Evergreens

Firethorn (*Pyracantha* spp.). Cuttings from current year's growth. Take in summer. Collect seeds, store in cool place, then stratify for 3 months at 40°F. before planting. Grafting possible on cotoneaster understock.

Forsythia (*Forsythia* spp.). Hardiest is *F. ovata*. Take hardwood cuttings in early spring, or softwood with leaves in late spring or summer. Division. Seeds unreliable.

Fothergilla (*Fothergilla major* or the dwarf *F. gardenii*). Seeds need 5 month stratification at room temperature and 3 months at 40°F. Softwood cuttings.

Fringe tree (*Chionanthus virginicus*). Difficult from cuttings. But softwood cuttings in late spring may succeed with indoleacetic acid, mist, and continuously sterile medium. Seeds need stratification for 1 month at room temperature and then 2 months at 40°F. before planting.

Gardenia—see House Plants

Ginkgo (*G. biloba*). Keep seeds moist after removing them from pulp in mid-autumn. Pack in damp sand for 3 to 4 months at 70°F. Then stratify at 40°F. for 3 to 4 months. Softwood cuttings from male trees. Avoid female trees because of their seeds' unpleasant odor.

Golden rain tree (*Koelreuteria paniculata*). Sow seeds when ripe, or keep 1 year in airtight container and then sow. Perforate seed with file, or give a 60-minute soaking in acid, followed by stratification for 3 months at 40°F. Softwood cuttings in spring. Root cuttings in spring.

Hackberry (*Celtis laevigata*). Sow seeds as soon as harvested, or soak in hot water to make removing pulp easier and then stratify for 2 to 3 months at 40°F. Cuttings more reliable. Grafting also possible.

Hawthorn (*Crataegus* spp.). Seeds held back by both seed coat and embryo conditions. Wash from fruit, stratify in moist peat moss for 3 to 4 months at 70°F. and then for 3 to 4 months at 40°F. Do not transplant unless very young; has taproot. Grafting. Attractive to birds.

Heath, tree (*Erica arborea*). Leafy, partially matured cuttings, in summer; easy, if put under glass and hormone powder used. Seeds grow well in greenhouse in winter. Use same methods for other heaths, for example, spring heath (*E. carnea*), twisted heath (*E. cinerea*), which needs trimming after flowering, blackeyed heath (*E. melanthera*), and the very hardy cross-leaf heath (*E. tetralix*). Bottom heat often helps.

Hemlock—see Evergreens

Hibiscus (*H. rosa-sinensis*). Cuttings from terminal shoots of partly matured wood, in spring or summer. Leaf-bud cuttings. Layering and air-layering recommended for those difficult to root. 'Single Scarlet' is easy to root; use it for rootstock for grafting the less vigorous cultivars. Whip grafting or side grafting, in spring.

Holly (*Ilex* spp.). Some seeds ready to plant at time of harvest; others need stratification of 3 to 5 months at room temperature and 3 months at 40°F. English holly (*I. aquifolium*) needs this, and will take 2 years to germinate. American holly (*I. opaca*) needs the same. Or cold winter storage in damp peat moss and sand, though in the southern part of its range, cold treatment not needed. All hollies easily grafted on dormant stock of American or English holly. Cuttings from semi-hardwood tips on current year's growth. Inkberry (*I. glabra*) very attractive to birds.

Honeylocust (*Gleditsia triacanthos*). Seeds have hard coats. Can be stored for several years at 32° to 45°F. Before planting, place seeds in water at 190°F. Sow those which swell since this indicates seed coat has been penetrated.

'Moraine' locust usually grafted on seedlings of the thorny type. Cuttings possible.

Honeysuckle (*Lonicera* spp.). Layering is simplest. Cuttings of softwood or hardwood easy also. Division. Seeds need stratification for 3 months at 40°F.—in fact some need warm, then cold stratification. If they do not germinate after planting, dig them up and stratify again. Attractive to birds.

Hydrangea (*Hydrangea* spp.). Softwood and hardwood cuttings, in spring. Division of smooth hydrangea (*H. arborescens*). Seeds sown when harvested, or stored a year in closed jars and then planted.

Inkberry—see Holly

Jacaranda (*Jacaranda mimosifolia*). Seeds of this beautiful tree are said to germinate and grow easily when taken ripe from the seedpod after blooming. Other methods not used.

Japanese quince (*Chaenomeles japonica, C. speciosa*). Clean seeds from fruit and plant in fall, or stratify for 2 to 3 months at 40°F. Division easy. Cuttings of semi-hardwood in late spring root well with mist. Root cuttings of 2 to 3 inches do well if planted in spring after cold storage over the winter.

Jasmine (*Jasminum* spp.). Sow seed when ripe, or store in cool place in tightly covered jar until time to plant. Cuttings root easily. Layering also successful.

Juniper—see Evergreens

Labrador tea (*Ledum groenlandicum*). Hardwood cuttings. Division in spring.

Leucothoe (*L. fontanesiana* or *L. catesbaei*). Softwood and hardwood cuttings easy. Sow seeds when ripe or store in cold for up to a year. Division of clumps in spring.

Lilac (*Syringa vulgaris* and cultivars). Cultivars do not come true from seed, but you may get some nice surprises if you plant outdoors in fall, or stratify for 2 months at 40°F. before sowing. Softwood cuttings must be taken just before, during, or just after lilacs flower. Cut shoots when 6 to 8 inches long, trim off lower leaves and prevent wilting with mist-spray. Grafting is easier; plant the union point well below ground level to encourage roots from scion. Cut back suckers. Use privet for understock because its suckers are easy to identify.

Linden tree (*Tilia cordata* et al.). The littleleaf linden, an excellent street tree, is well worth propagating by budding or grafting, or from seeds. Also try basswood (*T. americana*). Stratify seeds for 5 months at 35°F. after soaking in hot water, or removing the seed covering. If germination fails, dig up the seeds and stratify again. Viable for 10 to 15 years at low temperatures. Good for bees.

Locust, black (*Robinia pseudoacacia*). Give seeds double stratification of 5 months at warm room temperature and then 4 months at 40°F. Viable for 10 to 15 years at low temperatures. Budding and grafting on locust understock of same species. (See also honeylocust.)

Madrone, Pacific (*Arbutus menziesii*). Stratify seeds at 35° to 40°F. for 3 months. Difficult to transplant. Cuttings and layering also possible.

Magnolia (*Magnolia* spp.). Gather seeds from fruits in fall. Either plant at once, stratify 3 to 6 months or over winter at 40°F. for spring planting, or store dry at low temperature (40°F.). Softwood cuttings in late spring, but they need misting and bottom heat.

Mahonia (*Mahonia* spp.). Stratify seeds at 40°F. over the winter, never letting them dry out, except for seeds of red mahonia (*M. haematocarpa*), which will not be harmed. Plant seeds of Oregon grape (*M. aquifolium*) in mid-summer.

Maidenhair tree—see Ginkgo

Manzanita (*Arctostaphylos manzanita*). Layers root easily. Also cuttings.

Maple (*Acer* spp.). Dioecious plants that shed seeds either in spring or fall. Commonest spring shedders are sugar maple (*A. saccharum*) and red maple (*A. rubrum*). Gather seeds at once and plant at once before dry. Seeds of other maples should be kept moist and stratified for 4 months at 40°F. and then sown. Grafting and budding. Amur maple (*A. ginnala*) good for bees.

Metasequoia or dawn redwood (*M. glyptostroboides*). The tree discovered in China in 1944 by a University of California paleobotanist. It had been thought extinct. He wants people to try growing it in all corners of the country to help determine its adaptabilities and tolerances. So grow it from seed you can buy (which has been kept, presumably, in airtight containers in a cool place). It is viable for a year and probably longer, and no pretreatment is necessary. Once you get a metasequoia started you can make softwood and hardwood cuttings. Please notify University of California Botany Department at Berkeley of your successes.

Mock orange shrub (*Philadelphus* spp.). Softwood and hardwood cuttings of cultivars root easily. Division with an ax in early spring. Seeds need stratification at 40°F. for 2 months. Suckers root by themselves; move to make new plants.

Mountain andromeda (*Pieris floribunda*). Sow seed when ripe, or keep cold for a year and then sow. Cuttings

203

difficult, except for those of Japanese andromeda. (*Pieris japonica*).

Mountain ash (*Sorbus* spp.). Two cold stratification periods may be needed for seeds. Try one for 3 months at 40°F. and if seeds do not germinate, dig up and stratify again at 40°F. Collect seeds as soon as mature and remove from pulp. Cuttings not successful. Budding can be. Attractive to some birds.

Mountain laurel (*Kalmia latifolia*). Sow seeds as soon as ripe or stratify for 3 months at 40°F. Cuttings difficult.

Mulberry (*Morus alba, M. rubra*). Fruitless mulberries (with only male flowers) propagated by cuttings, budding, and grafting. Most successful are leafy softwood cuttings taken in mid-summer and put under mist. Berried trees yield seeds which need stratification for 3 months at 40°F. and are then sown. Very attractive to birds.

Oak (*Quercus* spp.). Do not allow acorns to dry out, and do not let squirrels go after them if you plant seeds in the fall. Cages help. White oak (*Q. alba* et al.) seeds need to be sown when ripe. Black oak (*Q. velutina* et al.) seeds need stratification for 3 months at 40°F. Grafting succeeds if stocks and scions are closely related.

Oleander—see House Plants

Olive (*Olea europea*). Cuttings root easily. Soak bottom 2 inches for 24 hours, then hold in sawdust until callus forms. Suckers can be moved and replanted. Seeds germinate slowly, but are used to grow seedlings for grafting. If planted, even pieces of trunk and bark will send up shoots.

Pear, Bradford (*Pyrus calleryana* 'Bradford'). An ornamental street tree. Root grafting, T-budding. Compatible with *P. calleryana*, but not *P. communis*. Leafy cuttings, in late summer, if mist and hormone powder used.

Peony, tree (*Paeonia suffruticosa*). To grow from seeds you must pay attention to the odd way they develop: the shoot tip of the embryo is still dormant after the roots begin to grow. You must plant the seeds in a moist medium in small pots, then move pots to cold room or refrigerator kept at 40° to 50°F., or put outdoors where that temperature will stay fairly constant. Leave in cold for 2 to 3 months to break dormancy of the root tip. Then return to warmth and the shoot tip will grow. Hybrids do not come true to seed, so use grafting for them, with garden peony for rootstock. After making graft, insert in sand and peat to permit callusing. Then plant in soil drawn up over graft union to encourage root growth from the scion.

Pine—see Evergreens

Plane tree and sycamore (*Platanus* spp.). Seeds must not dry out after collection in late winter. Plant at once or stratify for 2 months at 40°F. and plant in fall. Best to let seedballs winter on the tree. For storage of 1 year or longer put in airtight containers in the refrigerator. Softwood and hardwood cuttings sometimes successful.

Plumbago (*P. capensis*). Seeds sown in late winter. Root cuttings and semi-hardwood cuttings do best in moist growing conditions. Division.

Poplar, aspen, and cottonwood (*Populus* spp.). Seeds very short-lived. Plant as soon as ripe, or even earlier, as soon as capsules begin to open. No pretreatment requirements. Keep at low humidity, in closed jars, near to 32°F. if necessary; may live a year or so. Seedlings very susceptible to damping-off and to hot, dry winds. Hardwood cuttings, in spring. Softwood cuttings sometimes will root.

Privet (*Ligustrum* spp.). Seed propagation easy, if cleaned seed stratified for 2 to 3 months at 32°F. up to 52°F.

Hardwood cuttings in spring. Softwood in summer, under glass or polyethylene. *L. japonicum* not so easy as others; use terminal shoots.

Pussy willow, French (*Salix caprea*). Cuttings very easy, in spring. Root cuttings also. Seeds from capsules viable only a few days.

Quince, flowering (*Chaenomeles* spp.). Seeds easy; plant in fall or stratify 3 to 4 months at 40°F until spring. Leafy cuttings root well under mist; take in late spring. Well-rooted suckers easy to move.

Red bud (*Cercis canadensis*) **and Judas tree** (*Cercis siliquastrum*). Process seed as soon as ripe: soak in hot water at 190°F. and allow to cool in water overnight. Then stratify for 3 months at 40°F. Then sow. Can be stored in tight container before processing. Can sow in fall. Softwood cuttings, in early summer. Simple layering.

Rhododendron spp. Layering used to be the favored method, but now cuttings (under mist), grafting, and growing from seed are more widely used. High humidity needed for grafting as well as for cuttings. Use scion wood from current growth, place grafted plant in warm, moist environment. Then harden and plant out. To use seedlings of *R. ponticum* for rootstock, grow from seeds collected just when about to fall. Store seeds dry and in the cold until ready to sprinkle over sphagnum moss on base of moist sand and peat moss. Spray gently, use bottom heat, keep in shade. Keep seedlings cool the next winter and set out the following fall. Cuttings are much less complicated and arduous, so now are method most commonly practiced.

Rose of Sharon—see Hibiscus

Rose (*Rosa* spp.) Multiflora roses are not mentioned in this discussion; they are noxious weeds now and outlawed in some states.

APPROPRIATE METHODS

Softwood cuttings .. Miniatures, hybrid teas, floribundas

Hardwood cuttings .. Climbers, hybrid perpetuals, large bush

Division .... Climbers, bush, floribundas

Soil layering .... Ramblers, climbers, others

Tip layering .... Ramblers, others

Air layering .... Cane roses, climbers

Budding .... Hybrid teas, winter-hardy

Grafting .... Hybrid teas, grandiflora

Seeds .... For hybridizing, *R. rugosa* and other old bush roses.

Softwood cuttings: Take 6 inches from current year's growth when partly mature, from spring to late summer. Leave only top 2 leaves. Put outdoors in peat-perlite mixture under a jar; or indoors in well-drained peaty mixture. Can be left under jar until following year, or transferred to cold frame for the winter. Mist and bright light best for miniatures.

Hardwood cuttings: Take 6-inch cuttings in fall or early winter, soon after leaf-drop, using firm wood of previous season's growth, and cutting just below node. Store in bundles in moist sand, sawdust, or leaf mold in cool, damp, dark place. When cuttings have callus or roots by following spring, plant out with only top bud above soil surface.

Division: Best method for climbers or species having many stems below ground. Cut rose bush back to 10 inches, saw through plant, dig and pull roots apart. Replant 2 inches deeper than before, filling new holes with peat moss and sand if needed. Divide spring or fall; in fall use heavy, slightly acid mulch.

Soil layering: Layer when canes are hardened: June to September. With climbers, bury section near tip of cane, using canes which have not yet bloomed. But with bush roses, use canes that have flowered, using branch that comes from near base. Slit halfway through branch, just below node. Put in 6-inch hole filled with sphagnum moss. Prop slit open with stick or stone and take off nearby leaves. Bend cane into hole, anchor with largish stone and cover with 4 inches of soil. Add another stone, and make ditch to catch water if season is dry. Can also water by inserting tin can with hole punched in it; bury this 2 inches into soil near point of layering. When rooted and you transplant the new plant, prune to half-size.

Tip layering: Best method for ramblers, which do it naturally anyway. When new growth ripe in September, strip leaves off tip, and bend over the tip, pushing it 6 inches into the soil. Stake until rooted. Then cut parent cane 10 inches above ground, wait 1 week, and transplant.

Air layering: When first blooms gone in late spring, take new ripe cane and remove leaves and thorns for 5 to 6 inches at point 18 inches below tip. Cut with sharp knife or razor blade, force in sphagnum moss to keep cut open, and wrap with moistened sphagnum moss in 3-inch ball. When rooted, remove plastic wrap, but not moss. Cut cane, and plant in pot or in pot plunged in earth. (See chapter 3 on Layering.)

Budding: The common method of propagating roses. Use for stock, *R. canina*, *R. chinensis* 'Ragged Robin', or *R.c.* 'Gloire de Rosomanes'. Get budwood from desired cultivar from current season's growth. Clip off leaves at once, but leave petiole of ½ inch. Use plump, dormant buds from side of stems which produce flowers, and which have thorns still easy to snap off. Put buds into ½-inch wood of original cuttings, not new growth. Overwinter with bud just below ground level. (See chapter 5 on Budding and Grafting.)

Grafting: Same method and roses as for budding, but with 2 buds instead of 1. Rarely used.

Seeds: Used for creating new hybrids. Also for old-fashioned roses which will come true. Gather hips (rose fruits) when ripe, just before first frost. Mash and float out seeds. Stratify at once at 35° to 40°F. for 4 to 6 months for roses such as *Rosa rugosa* and *R. hugonis*. Leave seeds of *R. canina* in moist vermiculite at room temperature for 2 months; then stratify at 32°F. for 2 months. For hybrid teas, change stratification temperature to 36° to 40°F. Germination slow. Seedlings from fall planting should be 4 inches or more tall to survive winter in cold areas. Otherwise plant in spring, though you can leave seeds in large pot or flat outdoors all winter, on north side of house or garage where alternations of thaw and freeze are less likely than in warmer exposures. Rose hips are attractive to birds.

Russian-olive (*Elaeagnus angustifolia*) **and silverberry** (*E. argentea*). Cuttings of semi-hardwood or hardwood. Grafting. Seeds. Attractive to birds.

Salal (*Gaultheria shallon*). Division. Layering. Seeds in acid soil.

Sarsaparilla and the angelica trees (*Aralia* spp.). Seeds need double pretreatment; stratify first in moist peat moss at room temperature for 3 to 5 months (or scarify seeds), then stratify at 40°F. for 3 months. Root cuttings. Budding onto roots. Wear gloves to avoid dermatitis from the southern aralia, devil's walking stick (*A. spinosa*).

Sassafras (*S. albidum*). Root cuttings in early spring. Seeds, but stratify for 3 months at 40°F.

Shadblow or serviceberry (*Amelanchier alnifolia*). Leafy softwood cuttings easy; take when new growth is 3 to 4 inches long. Use mist. Root cuttings. Embryo dormancy in seeds calls for 3 to 6 months' stratification at 35°F. Attractive to birds.

Siberian pea tree (*Caragana arborescens*). Softwood and hardwood cuttings. Seeds can be sown as soon as ripe, or dried and stored in airtight containers, or stratified for 2 weeks at 40°F. in moist sand. Germination improved by soaking seeds in water at 190°F. overnight. Attractive to bees.

Silktree (*Albizzia julibrissia*). Seeds need nicking and hot water pretreatment to break very tough seed coat. Root cuttings; use ½ inch in diameter and 3 to 4 inches long. Stem cuttings unsuccessful.

Skimmia (*S. japonica*). Collect fruits from female plants; soak to remove seed from pulp. Plant when cleaned. Stem cuttings.

Smoke tree (*Cotinus coggygria*). Softwood cuttings, but difficult. Layering. Seeds need stratifying for 3 months at room temperature, then for 3 months at 40°F. to break double dormancy, but asexual methods much better.

Spirea (*Spiraea prunifolia, Spiraea* x *vanhouttei* et al.). Best propagated by cuttings, either leafy softwood, taken in summer, or hardwood, taken in spring (especially good for Vanhoutte spirea).

Spruce—see Evergreens

Tamarisk (*Tamarix* spp.). Cuttings easy, either softwood or hardwood. Prunings of *T. parviflora* after flowering, or of *T. pentandra* before flowering, when still dormant in very early spring, may be used. With seeds, no pretreatment is necessary.

Tree-of-heaven (*Ailanthus altissima*). This dioecious species self-seeds freely when trees of both sexes are nearby. To plant seeds yourself, embryo dormancy requires stratification for 2 months at 40°F. Root cuttings, but use female trees because male staminate trees have very offensive odor.

Tulip tree (*Liriodendron tulipifera*). Seeds should not dry out. Stratify for 2 months, varying the temperature daily between 32°F. and 50°F. Or try a constant 40°F. for 2 to 4 months. Or plant outdoors in fall. Germination poor; many seeds lack embryos. Leafy stem cuttings, taken in summer. Root cuttings. Do not transplant unless you grow in containers or young plants are specially balled and burlapped.

Tupelo (*Nyssa salvatica*). Seeds must be kept moist. When they are ripe, stratify for 3 months at 40°F. Suckers and rooted layers can be removed to make new plants.

Viburnums (*Viburnum* spp.). There are many varieties. Most require double stratification of 5 months at room temperature and then 3 months at 40°F. A few varieties require 9 months at room temperature, yet for *V. sieboldi,* a Japanese viburnum, a single cold stratification is enough. Seed can be stored dry for up to a year. Softwood and hardwood cuttings. Division. Grafting. Layering. Those methods all possible with *V. odoratissimum, V. davidii, V. suspensum,* and *V. tinus. V. dentatum* very attractive to birds.

Weigela (*Weigela* spp.). Hardwood cuttings, early spring. Softwood cuttings late spring to fall. Put under glass or polyethylene.

Willow (*Salix* spp.). Stem cuttings very easy. Root cuttings very easy. Seeds difficult; you must gather as soon as capsule opens, plant immediately. Viable only 3 to 4 days, or with exactly proper moisture and degree of cold for 4 to 6 weeks only. Germination in 12 to 24 hours. Attractive to bees in early spring.

Wintergreen (*Gaultheria procumbens*). Tiny shrub. Layering. Division. Seeds, which need 3 months stratification at 40°F. You can store seeds for a year in a cool, dry place at 40°F. Use airtight container.

Witch hazel (*Hamamelis* spp.). Layering. Grafting, using *H. virginiana* as rootstock. Seeds require double stratification: 5 months at room temperature and 3 months at 40°F. Softwood cuttings also possible.

Yew—see Evergreens

Most of the annuals listed here are propagated from seeds, including the perennials grown as annuals to bloom the first season. Many seeds can be started indoors in early spring, especially if the plants are tender. Seeds for perennials, however, can be started from May to September outdoors and from February on indoors.

For germination times see the information on seed packets or any of the excellent USDA lists and pamphlets on growing plants. Or your favorite gardening book. Helpful bottom heat to use for seeds and cuttings in flats runs from 60° to 85°F., with an average of around 70°F. Seeds are often viable for 3 to 10 years, so do not throw them out until you have tested them for germination. Remember not to plant seeds too deep and to keep young, tender, or disturbed plants properly watered, aired, and shaded if they show signs of wilting.

Garden Flowers

Aaron's rod (*Thermopsis caroliniana*). Perennial. Also called bush pea or golden pea. Sow seeds when fresh in late summer. Division in spring (not always successful).

Achimenes spp. Tender bulbs. Propagate by offsets, division, or by seeds.

Adam's needle—see Yucca filimentosa

African daisy (*Arctotis grandis*). Annual. Start seeds indoors in April; or sow outdoors in May or when ground is warm.

African iris (*Moraea iridioides*). Tender bulb. Propagate by taking offsets from corms.

Ageratum (*A. eupatorium*), perennial; (*A. houstonianum*), annual. Sow seeds indoors February or March, outdoors in May. Takes 4 months to blossom. Do not cover seeds. Divide perennials yearly in spring. Cuttings. Pot up for house plants in fall, or sow seeds in fall for house plants.

Alkanet (*Anchusa italica*). Perennial. Before planting seeds, refrigerate 48 hours. Set outdoors anytime during summer, in semishade. Root cuttings spring or fall. Division spring or fall.

Allium spp. Hardy bulbs of onion family. Range from small *Allium moly,* with yellow blooms, to big lavender-blooming *A. gigantea.* Lift and divide in summer when dormant. Some have aerial bulblets; plant when ripe. Sow seeds when ripe. Used as pest-repellents.

Alyssum, basket of gold (*Alyssum saxatile* or *Aurinia saxatile*). Spring-flowering perennial. Sow seeds spring or early fall; thin to permanent position. Also lift, pot up for winter. Division, if it lives that long.

Alyssum, sweet (*Lobularia maritima* or *A. maritima*). Annual. Start indoors in March, or outdoors where it is to grow. If fresh seed used, prechill it. Sometimes self-sows. Plant in succession for longer season of bloom. Try planting in fall, too.

Amazon lily (*Eucharis grandiflora*). Tender bulb. Patio plant. Divide when repotting. Remove offsets.

Amsonia (*Amsonia taberaemontana*). Perennial. Sow early spring. Divide early spring.

Anemone spp. Perennials. Divide fibrous-rooted *A. japonica* (but they don't really like it). Or sow seed in February or March (but they probably will never come true). Soak seeds of pasque flower (*A. pulsatilla*) for 6 to 8 hours before planting. (See also Pasque flower in Wild Flowers.) Lift tubers of bulbous anemones in June or July; replant in fall.

Angel's trumpet—see Datura

Aster spp. Perennials. Michaelmas daisy (*A. novae-angliae*) is hybrid. Does not come true from seeds. Divide every 2 to 3 years, discarding old central growth. Sow non-hybrid seed varieties in February indoors. Sow seed of dwarf aster (*A. alpinus*) in early spring. Divide every 3 years. See also China aster.

Astilbe (*A. simplicifolia alba, A. s. rosea*). Perennial. Also called false spirea, especially by florists. Sow seeds in early spring. Divide spring or fall.

Autumn crocus (*Colchicum autumnale*). Hardy bulb. Also called meadow-saffron. Lift bulbs in June and replant. Plant seeds in August or September in cool greenhouse. Matures in 5 years.

Avens (*Geum* spp.). Perennial. Try 'Scarlet Double', sowing seeds in February or March. Divide every 3 years in spring or fall. Save seeds of wild avens in fall; sow in spring.

Aztec lily (*Sprekelia formosissima*). Tender bulb. Lift in fall, store with tops on; plant when soil warm in spring. Or use as patio plant, lifting bulbs for rest period.

Baby blue eyes (*Nemophila insignis* or *N. menziesii*). California perennial, used as annual in North. Sow seeds outdoors in April. Plant in full sun. Winter-kills except in South, or warm parts of California et al. Self-sows.

Baby's breath (*Gypsophila elegans*), annual; (*G. paniculata*), perennial. Plant seed of annual outdoors fall or mid-April. Thin. Often self-sows. Sow perennial seeds indoors in February or March; outdoors later. Cuttings spring or fall. Root cuttings. Grafting.

Bachelor's-button (*Centaurea cyanus*), annual; (*C. dealbata*), perennial. Sow annual seeds in March indoors;

after soil warm, outdoors. Often self-sows. Divide perennial every year or so. Cuttings in September to hold over indoors where cool. Sow seeds indoors in March.

Balloon flower (*Platycodon grandiflorum*). Perennial. Sow in spring, or even better in a cold frame in fall when seeds ripe. Division very difficult.

Balsam—see Patience

Baneberry—see Wild Flowers

Barren strawberry—see Wild Flowers

Basket flower (*Centaurea americana*). Annual. Plant early outdoors; grows very slowly.

Basket of gold—see Alyssum

Beach wormwood—see Wild Flowers

Bearberry—see Wild Flowers

Beard tongue (*Penstemon* spp.). Perennial. Sow in warmth indoors; outdoors in late summer. Divide early spring every few years. Take cuttings from side shoots to winter over and plant out in spring. *Penstemon* x *gloxiniodes* is perennial treated as annual in North. Sow fall or spring.

Bear's breech (*Acanthus mollis*). Perennial. Sow in spring or fall. Rather difficult from seeds. Divide spring or fall.

Begonia, tuberous—see Tuberous Begonia

Bellflower (*Adenophora communis*). Perennial. Also called ladybell. Sow in spring where to grow. Resents moving or division. Cuttings.

Bellflower (*Campanula* spp.). Biennial, perennial, annual. From seed, some easy, some difficult. Sow spring to September; do not cover seeds. Division usually spring. Sow *C. ramosissima,* the common annual, in early March indoors or in cold frame. Also called bluestar bellflower.

Bells-of-Ireland (*Molucella laevis*). Annual. Germinates only where warm; slow outdoors. Self-sows, but no germination until warm weather.

Bergenia (*Bergenia cordifolia*). Perennial. Also called saxifrage. Divide spring or fall. Sow seeds in spring.

Betony (*Stachys lanata*). Perennial. Also called lamb's ears. Sow indoors in February or March. Divide every 3 to 4 years.

Bishop's cap—see Wild Flowers

Black cohosh—see Wild Flowers

Black-eyed Susan—see Wild Flowers

Blanket flower (*Gaillardia* spp.). Annual, perennial. Sow seeds outdoors in April, or indoors in March. Root cuttings: cover with ¼-inch pure sand and press down. Stem cuttings, later in summer. Use side shoots. Self-sows.

Bleeding heart (*Dicentra spectabilis*). Perennial. Do not move, but take stem cuttings, root cuttings, or divide without moving. Stem cuttings should come from base of flower clusters, after flowering. Root cuttings same time, 3 inches long and plant 3 inches deep. Mulch in winter. Stratify seeds at 40°F. over winter or sow in fall. Fringed bleeding heart (*D. eximia*) much less fussy about being moved; see Bleeding heart in Wild Flowers.

Blood-lily (*Haemanthus multiflorus*). Tender bulb. Lift and take offsets, bulblets, after flowering in the fall. Does best in greenhouse or Zones 9 and 10.

Bloodroot—see Wild Flowers

Blue cohosh—see Wild Flowers

Blue-dicks (*Brodiaea capitata*). Tender bulb of Oregon. Divide bulblets.

Blue wild indigo (*Baptisia australis*). Perennial. Collect seeds from capsule in early fall, sow in spring. Divide spring or fall. Do not transplant at any other time. Self-sows.

Blue wings (*Torenia fournieri*). Annual. Sow indoors in March; needs warmth of 70° to 75°F. Long germination, but rapid growth. Self-sows. Pot up in fall for house plant.

Bluestar bellflower—see Bellflower

Boneset—see Wild Flowers

Bouncing Bet (*Saponaria officinalis*). Perennial. Divide in spring. Stem cuttings in summer. Sow early indoors or when warm outdoors. (See also Soapwort.)

Browallia (*B. americana, B. speciosa*). Annual. Easy from seed, indoors or out, and if in greenhouse, plant as early as February. Pot up for house plant; it can bloom all winter.

Bugbane (*Cimicifuga racemosa*). Perennial. Also called snakeroot. Divide spring or fall. Sow seed when ripe.

Bugloss, blue bird (*Anchusa capensis*). Annual. Sow indoors early, outdoors in May.

Bunchberry—see Wild Flowers

Bunchflower—see Wild Flowers

Burnet—see Wild Flowers

Buttercup (*Ranunculus* spp.). Perennials. Either fibrous or tuberous rooted. Divide clumps in spring. Seed sown when ripe and left in pot for a year then can be set out to bloom the following summer. *R. asiaticus* has tuberous roots; to be lifted before frost and dried before replanting in the winter in the greenhouse. Plant with claws down, not up. Tender.

Butterfly flower—see Poor man's orchid

Butterfly weed (*Asclepias tuberosa*). Perennial. Division. (May be poisonous to animals.)

Caladium (*Caladium bicolor*). Tender tuberous plants. In South plant out-of-doors; in North take up each fall, divide, and store until replanting June 15, or earlier in the greenhouse.

Calendula—see Pot-marigold

California bluebell (*Eutoca californica*). Annual. Plant seeds indoors in March, in 70°F. environment. Outdoors when warm.

California poppy (*Eschscholtzia californica*). Annual. Plant outdoors in March or September. Do not transplant. Often self-sows.

Calla palustris—see Wild Flowers

Calliopsis (*Coreopsis drummondi*). Annual. Sow indoors in March. Or outdoors May or September. Self-sows.

Campion, red—see Wild Flowers

Candytuft (*Iberis affinis, I. amara, I. umbellata*), annuals; (*I. sempervirens* et al.) Perennials. Easy if seeds of annuals sown in spring where plants are to grow. For perennials, softwood and hardwood cuttings. Seeds sown as soon as ripe. Perennials easily divided.

Cape marigold (*Dimorphotheca annua*). Annual. Can start seeds indoors, but outdoors where they are to grow is better. In warm climates, plant in winter for spring bloom.

Cardinal flower—see Lobelia; see also Wild Flowers

Carnation (*Dianthus caryophyllus*). Tender perennial, greenhouse plant also called clove pink. Cuttings November or December, to plant out after frost. Sow seeds late spring where climate's mild; in greenhouse at 60°F. in the day and 50°F. nights. Use bottom heat for both seeds and cuttings.

Castor bean (*Ricinus* spp.). Annual. Big, mole-repellent plant. Sow seeds outdoors after frosts. Or indoors in April, after soaking and putting in Jiffy pots®. Seeds poisonous.

Celandine (*Chelidonium majus*). Perennial. Shake seeds from plant and sow in September. Self-sows. Divide in spring. Take root cuttings in summer.

Chamomile—see Herbs

China aster (*Callistephus chinensis*). Annual. Indoors seeds need bottom heat of 70°F. Fresh seeds need bottom heat of 86°F. Lights for 12 hours encourage bloom. Outdoor sowing after frost, for later bloom. Do not put near pot-marigolds.

Chinese forget-me-not (*Cynoglossum amabile*). Annual or biennial. Also called hound's tongue. Sow outdoors where to grow. Self-sows freely. Indoors use lights.

Chinese lantern (*Physalis alkekengi*). Perennial. Division and seed. Use lights. Root cuttings.

Christmas rose (*Helleborus niger*). Perennial. Sow fresh seeds in cold frame. Brittle roots need great care during division in August or September.

Chrysanthemum spp. Annual, perennial. Seeds of the annual *C. carinatum* have very low germination rate, but try both indoor and outdoor planting. Cuttings possible. Perennial species need yearly propagation by stem cuttings. Division every 2 to 3 years.

Cinquefoil—see Wild Flowers

Clarkia (*Clarkia elegans*). Annual. Plant indoors in light and coolness of 55°F. Or plant outdoors in fall or very early.

Clover (*Trifolium repens*). Perennial. Seeds only.

Cockscomb (*Celosia cristata, C. plumosa*). Annual. Start seeds indoors in April with bottom heat; or outdoors when frost no longer a threat. Germination slow outdoors. Use Jiffy pots® if you are going to transplant, and fertilize well with compost tea or manure tea.

Coleus (*C. blumei*). Annual. Start indoors in February, at 70°F.

Columbine (*Aquilegia* spp., *A. caerulea*). Perennial. Sow outdoors in August, or indoors in early spring. Chill seeds for a month before planting. Self-sows. Divide early spring or late summer. See also Wild Flowers.

Coneflower (*Rudbeckia subtomentosa* et al.), perennial; (*R. bicolor*), annual. Divide every few years. Sow early spring in cold frame. Self-sows. Sow annual seeds in Jiffy pots® in February or March. Barely cover the seed.

Coral-bells (*Heuchera sanguinea*). Perennial. Prechill seed for 4 weeks in cold frame or refrigerator. Then plant inside in February or early March. Will bloom following year. Divide every 3 years. Stem cuttings or leaf cuttings with heel will root easily.

Coreopsis (*Coreopsis grandiflora*). Perennial. Also called tickseed. Indoors seed will germinate in 4 to 5 days; or sow outdoors in cold frame or where to grow. Resow every year for maximum bloom. Divide fall or spring. Used as pest-repellent.

Corn lily (*Ixia* spp.). Tender bulb. Divide corms in late summer. Store and plant in spring.

Corydalis (*Corydalis lutea*). Annual or perennial. Sow seed as soon as ripe. Divide or move only in spring. (See also Wild Flowers.)

Cosmos spp. Annuals and perennials. Plant outdoors where to grow. Give high (85°F.) heat if indoors.

Crane's bill (*Geranium maculatum*). Perennial. Divide in spring. Sow seeds in spring.

Crimson flag (*Schizostylis coccinea*). Tender rhizomatous plant. Divide rhizomes when you lift plant for the winter to store or pot up for winter bloom.

Creeping zinnia (*Sanvitalia procumbens flore-pleno*). Annual. Plant indoors in March or very early outdoors. Likes heat and sun. *Crocus* spp. Cormous plant. Division of cormels from corms, but do it rarely. (See also Autumn crocus.)

Cupflower (*Nierembergia caerulea*). Tender perennial grown as annual. Sow seeds indoors in February in 70° to 80°F. warmth. Cuttings possible.

Cupid's blue dart (*Catananche caerulea*). Annual. Plant indoors in March or outdoors when soil workable. Set out plants after frost.

Cyclamen, hardy (*Cyclamen europaeum*). Hardy cormous plant. Sow seeds in August. Matures in 1½ years; do not crowd corms. Lift to divide when necessary in July. Replant then. (See also House Plants.)

Daffodil—see Narcissus

Dahlia (*Dahlia* in variety). Tender tuberous plant. Separate tubers at planting time in spring, leaving one good strong eye and a part of stem on each tuber. Start nonhybrids from seeds indoors in spring, then pot and keep in garden during summer. Stem cuttings of small shoots from base of plant, but do not cut off all the shoot. Use bottom heat for 3 weeks to get the roots started. Pot and force to be pot-bound by fall. Tubers will form; use them for propagation, too.

Daisy, painted (*Chrysanthemum coccineum*). Perennial. Also called pyrethrum. Divide late summer. Sow seeds in spring at 68°F.

Daisy, Shasta (*Chrysanthemum maximum*). Perennial. Easy to grow from seed sown in spring. Divide stolons in spring.

Datura or angel's trumpet (*Datura chlorantha*). Annual. Sow in late March indoors at 70°F. or higher. Germination may be slow. Outdoors plant in May when warm. Even larger and more handsome is *D. metel,* which often self-sows. (In hot climates, called Jimson weed. Poisonous.) Used as a pest repellent.

Daylily (*Hemerocallis fulva, H. flava, H. aurantica* et al.). Perennial. Division of clump or rootstock. Seed of those which produce seed sown in spring; matures in 2 to 3 years. Pollen of *H. aurantica* varieties used for breeding, as is that of *H. nana, H. plicata,* or modern cultivars, especially including *H. fulva* 'Europa'.

Delphinium (*D. grandiflorum*). Perennial. Propagate from seeds you save from a few stalks not cut back after blooming. Sow late July or August, or hold over to February at 40°F. in refrigerator, and start indoors. Cover seeds very lightly, with fine sand or sphagnum moss. For annual delphinium, see Larkspur.

Dianthus (*Dianthus chinensis*). Annual. Also called pinks. Sow in cold frame or outdoors when soil is warm. Save own seed and stratify in refrigerator, at 42°F. for 3 weeks before sowing.

Dusty miller (*Centaurea* spp.). Annual and perennial. Perennials *C. cineraria* and *C. gymnocarpa* by seed, cuttings, division, all easy. Start annuals like bachelor's-button, *C. cyanus,* indoors with slight covering of sphagnum moss at 70°F., move to flat and put outdoors to harden before planting in garden. (See also Bachelor's-button, and beach wormwood in Wild Flowers.)

Evening primrose (*Oenothera drummondi*), annual; (*O. fruticosa*), perennial. Plant annual seeds in Jiffy® pots indoors in March; resents transplanting. Or sow outdoors in May. Start perennial seeds in February or March. Divide in spring after blooming or in late summer, every 2 to 3 years.

Everlasting (*Helipterum manglesii*). Annual. Sow outdoors where plants are to grow. Also called sunray.

False dragonhead (*Physostegia grandiflora*). Perennial. Also called obedient plant. Easy to grow from seed. Divide creeping stolons in spring every year.

False spirea—see Astilbe

False starwort (*Boltonia asteroides*). Perennial. Sow in spring. Divide in spring every 3 to 4 years.

Farewell-to-spring (*Godetia amoina, G. grandiflora*). Annual. Plant in late spring outdoors where they are to grow.

Faun lily (*Erythronium* spp.). Bulb. Also called trout lily. Divide in fall. Dig deep and put small stone under replanted bulblets to keep them from burrowing.

Ferns—see Ferns

Feverfew (*Chrysanthemum parthenium*). Perennial. Sow seed in February or March. Divide in March. Stem cuttings, with heel, from young shoots at base of mature plants.

Firecracker plant (*Cuphea ignea*). Annual. Also called cigarflower. Long growing period so plant indoors in March.

Flax, flowering (*Linum grandiflorum*), annual; (*L. perenne*), perennial. Plant annual seeds outdoors in successive plantings where they are to grow. Divide perennials spring or fall. Sow seeds when ripe, or in spring. Stem cuttings in late summer. Use limey soil.

Fleabane (*Erigeron* spp.). Perennial. Divide in spring. Sow seeds in spring.

Flowering tobacco (*Nicotiana alata*). Annual. Sow seeds indoors in April or outdoors in late spring. Do not cover seeds, which are very small—just press them into soil or sphagnum moss. Only cloudy days are suitable for transplanting. Pot up for winter house plant. Often self-sows.

Foamflower—see Wild Flowers

Forget-me-not (*Myosotis dissitiflora*). Biennial; (*M. scorpioides*), perennial. Biennial is no bother, it self-sows abundantly. Perennial sown indoors in March or outdoors in mid-July. Self-sows readily. Cuttings root easily in wet sand. Divide in spring.

Four-o'clock (*Mirabilis jalapa*). Annual. Also called marvel-of-Peru. Lift in fall, store tuberous roots in frost-free place; plant out in May. Sow seeds indoors in March; outdoors in May or when soil warm. In moist shade, it self-sows.

Foxglove (*Digitalis grandiflora*). Biennial or perennial. Usually short-lived, so sow seeds each spring outdoors in May. Also divide in early spring. *D. purpurea,* biennial: sow seeds in spring or summer.

Foxtail lily (*Eremurus* spp.). Tender tuberous perennial. Also called desert candle. Sow seed in cold frame as soon as ripe. Best not to divide, but you can (not very successfully), and replant with roots 4 to 6 inches below surface of ground. Mulch with wood ashes, coal ashes, or sand. The hardy one is *E. himalaicus.*

Fraise des bois (*Fragaria vesca*). Perennial. Also called alpine strawberry. Sow seeds early spring on top of rich soil. Divide early spring.

Fritillary (*Fritillaria meleagris*). Hardy bulb. *F. imperialis,* the crown imperial, also fairly hardy. Lift and separate offset bulblets every 2 to 3 years. Used as a mole-repellent.

Funkia (*Hosta plantaginea*). Perennial. Also called plantain lily. Easy to grow from seed, as are all the funkias. Divide in spring.

Gaillardia (*G. pulchella*). Annual. Plant in garden when soil warm, where it is to grow. Stem cuttings. Often self-sows.

Galax—see Wild Flowers

Gamolepis tagetes. Annual. You may have to send to England for seeds. Sow them indoors in March; outdoors after frosts.

Garden heliotrope—see Valerian

Gasplant (*Dictamus albus*). Perennial. Also called dittany. Stem cuttings after flowering. Root cuttings of about 3 inches, sandy soil. Sow ripe seeds; or prechill for 6 weeks at 40°F. and then pour on boiling water to break dormancy. Can cause rash on people allergic to gasplant.

Gayfeather (*Liatris spicata*). Perennial. Stem cuttings. Divide in spring. Sow seeds in spring.

Gazania (*Gazania splendens*). Annual. Seeds must be sown indoors, in February or March. Set out in full sun.

Gentian—see Wild Flowers

Germander (*Teucrium lucidum*). Perennial. Divide spring or fall. Sow seeds when ripe. Softwood cuttings in early summer.

Gladiolus (*Gladiolus* hybrids). Tender corms. Divide corms and cormels when lifted in fall for storage. Or in South, lift, divide, and return to soil in late fall. Mulch.

Globe amaranth (*Gomphrena globosa*). Annual. Start indoors in March at 68°F. and keep cool. Or sow outdoors in late spring.

Globeflower (*Trollius europaeus*). Perennial. Divide just after flowering. Sow seed as soon as ripe.

Globe thistle (*Echinops exaltatus*). Perennial. Division. Root cuttings. Spreads easily where shady.

Gloxinia—see House Plants

Goatsbeard (*Aruncus sylvester*). Perennial. Divide in spring. Easy to grow from seed sown in spring.

Golden club—see Wild Flowers

Goldencup (*Hunnemannia fumariaefolia*). Annual. Plant seeds where they are to grow in the garden in spring. Or start in pots indoors in April.

Golden daisy (*Doronicum caucasicum*). Perennial. Also called leopardbane. Seeds may be erratic. Divide in late summer to get more reliable results.

Goldenrod (*Solidago* spp.). Perennial. Sow in spring. Divide in spring or fall.

Golden spider lily (*Lycoris* spp.). Hardy bulb. Remove bulblets and offsets.

Grape hyacinth (*Muscari* spp.). Hardy bulb. Divide clumps in July and separate bulblets. Sow seeds in spring, but those of white flowers may revert to species blue in seedlings. Self-sows. Flowering in 3 to 4 years.

Grass, ornamental. Sow seeds in spring. Hand-pollinate if you wish; with *Zea mays*, mix with pollen of other corns for mixed colors. Consult your county agent or a good garden encyclopedia for perennial grasses appropriate to your climate. Divide clumps.

Big quaking grass (*Briza maxima*). Annual.
Bristly foxtail grass (*Setaria verticillata*). Annual.
Cloud bent grass (*Agrostis nebulosa*). Annual.
Golden-top grass (*Lamarckia aurea*). Annual.
Hare's tail grass (*Lagurus ovatus*). Annual.
Japanese love grass (*Eragrostis tenella*). Annual.
Little quaking grass (*Briza minor*). Annual.
Rainbow grass (*Zea mays japonica*). Annual. Also called corn.
Ruby grass (*Tricholaena rosea*). Annual.

Groundsel (*Senecio aureus*). Perennial. Germinate seeds at 68° to 70°F. Keep very wet. Division. Stem cuttings. Root cuttings. *S. elegans:* start seeds in February. (See also Wild Flowers.)

Guernsey-lily (*Nerine sarniensis*). Tender bulb plant. Lift and divide bulblets from mother bulbs in late summer. Replant at once. Good for pots and tubs. Move indoors when frosts come.

Harebell phacelia (*Phacelia campanularia*). Annual. Plant indoors or outdoors, in mid-April, though germination time can take up to 3 weeks.

Heaths (*Erica carnea* et al.). Perennial. Many heaths best propagated by fall division and fall stem cuttings, but sow *E. carnea* early spring or fall.

Heathers (*Calluna vulgaris* et al.). Hardier perennials than heaths. Propagate the same way.

Heliotrope (*Heliotropium europaeum*). Perennial. Start seeds indoors in March with warmth of 77° to 80°F. as long as they do not grow spindly. If they do, reduce heat. Outdoors, seeds take 4 to 5 times as long to germinate.

Hens-and-chickens (*Sempervivum tectorum* et al.). Perennial. All easy by offsets. Or sow seed from early spring to early summer.

Hollyhock (*Althaea rosea*). Biennial, but 'Indian Spring' is annual. Self-sows. Start seed indoors in March; outdoors in April.

Honesty (*Lunaria annua*). Annual or biennial. Self-sows. Sow seeds in May for bloom the following year. Can save your own seeds, gathered when you remove outer seedpod membranes in the late summer or fall. Will be viable for 2 years.

Ice plant (*Mesembryanthemum criniflorum*). Tender perennial or annual. This so-called stone-similar plant grown in California, and in North as annual, where you sow seeds in February or March, and transplant after all danger of frost is gone.

Impatiens—see Patience

Indian cucumber root—see Wild Flowers

Iris spp. Perennials. For rhizomatous types such as the bearded iris, divide rhizomes in late summer. For bulbous ones, remove offsets, the little bulblets. Yellow iris (*I. pseudocorus*), also called yellow flag, will self-sow freely once established.

Ironweed—see Wild Flowers

Ismene—see Spider lily

Jack-in-the-pulpit—see Wild Flowers

Jacob's ladder (*Polemonium caeruleum*). Perennial. Sow spring or fall. (See also Wild Flowers.)

Joe-pye-weed—see Wild Flowers

Johnny-jump-up (*Viola tricolor*). Short-lived perennial. Also called heartsease, love-in-idleness, and viola. Self-sows abundantly. Division. Sow seeds indoors in January in the light, on top of soil. Or start in shade outdoors in August. Save your own seeds. Take cuttings in June or July for propagating hybrids.

Kafir-lily (*Clivia miniata*). Tender bulb. Divide bulbs in fall. Seeds take 5 to 7 years to grow to blooming stage.

Kenilworth ivy (*Cymbalaria muralis*). Perennial. Sow in the spring. Division. Stem cuttings.

Kingfisher daisy (*Felicia bergeriana*). Annual. Sow indoors in March, or in cold frame in April, or in garden in May. Best germination at 70°F.

Kochia—see Summer cypress

Ladybell—see Bellflower

Lamb's ears—see Betony

Lantana (*Lantana camara*). Tender perennial treated as annual in the North. Propagation by stem cuttings, but you can start your own from seed sown in January.

Larkspur (*Delphinium ajacis*). The annual delphinium. Sow late fall or spring. Often self-sows. Save seed and stratify at 40°F. for 3 weeks before sowing. Keep cool, at 45° to 55° F. Stem cuttings.

Leadwort (*Ceratostigma plumbaginoides*). Perennial. Also called plumbago. Divide spring or fall. Stem cuttings.

Lily (*Lilium* spp.). Bulbous plant. Divide scales from bulbs; take bulblets and bulbils. See chapter 4.

Lily-of-the-Nile (*Agapanthus orientalis*). Tender bulb. Dry out and divide bulbs in late fall. Store. Replant after all danger of frost has gone. Will survive in climates where temperature never goes below 30°F.

Lily-of-the-valley (*Convallaria majalis*). Rhizomatous perennial. Divide rhizomes spring or fall. Plant seed spring or fall.

Lobelia spp. Perennials. *L. cardinalis,* the cardinal flower, can be sown thinly spring or fall. Cuttings in midsummer. (See also Cardinal flower in Wild Flowers.) *L. erinus,* blue lobelia, can also be started early. Since lobelias prefer to germinate in the light, cover very lightly with sphagnum moss, or just press seeds down in. Regarded as annual because it winter-kills if not very carefully mulched and protected. Sometimes self-sows. Sometimes has offshoots.

Loosestrife (*Lythrum salicaria*). Perennial. Divide every few years. Stem cuttings. Sow seed indoors early spring; outdoors later.

Love-in-a-mist (*Nigella damascena*). Annual. Needs to grow where it is warm, whether grown indoors or out. Sow in fall in sandy soil where climate is warm; in North sow in spring. Will sometimes self-sow. Use successive plantings; it has short blooming period.

Love-lies-bleeding (*Amaranthus caudatus*). Annual. Sow indoors in April at 70° to 80°F. Outdoors in May, in full sun and soil not rich. Self-sows when conditions are right.

Lupine (*Lupinus* spp.). Annual and perennial. Difficult to transplant, so sow where to grow or use peat pots. Sow annual seeds indoors or outdoors in spring. Scarify perennial seeds; wait 3 to 4 days and soak 24 hours before sowing in pots. Self-sows. Cuttings from side shoots of hardened stems in spring. Resents moving.

Mallow (*Hibiscus moscheutos*). Perennial. Sow seed when ripe, but it can be stored 3 to 4 years. Once established, self-sows. Hardwood and softwood cuttings root easily. Layering. Grafting.

Marguerite, golden (*Anthemis tinctoria*). Perennial. Also called anthemis daisy. Easy from seed. Self-sows. Division. Stem cuttings.

Marigold (*Tagetes erecta, T. patula, T. signata pumila*). Annual. Plant seeds indoors in April, germinates in 5 days. Or plant outdoors when soil warms up. Easy to grow and transplant, but use red pepper to repel slugs when marigolds first put out. Sometimes they self-sow. Cuttings easy; they root in water. Try to grow *T. minuta* if you can. Used as a nematode-repellent.

Marsh marigold—see Wild Flowers

Maskflower (*Alonsoa acutifolia*). Annual. Start indoors in early April. Or outdoors in May after frosts.

Mayapple—see Wild Flowers

Meadow-beauty—see Wild Flowers

Meadow-foam (*Limnanthes douglasi*). Annual. Plant outdoors where to grow in May after frosts. Moist but sunny location.

Meadow-rue—see Wild Flowers

Meadow-saffron—see Autumn crocus

Merrybells—see Wild Flowers

Mexican shell flower (*Tigridia pavonia*). Cormous plant, tender in cold climates. Divide in fall after lifting. See Corms in chapter 4.

Michaelmas daisy—see Aster

Mignonette (*Reseda odorata*). Annual. Sow indoors in April, and it will germinate in 3 days, but it takes long to bloom.

Mistflower (*Eupatorium coelestinum*). Perennial. Divide in spring. Self-sows. Or save seed and plant indoors in February. Has stolons, so it spreads. *E. rugosum* called white snakeroot.

Monkey flower (*Mimulus cupreus*). Annual. Start seed indoors or in the greenhouse in March; plant out in shade after frosts. Cuttings also possible.

Monkshood (*Aconitum napellus*). Perennial. Also called wolfbane. Chill seeds at 40°F. for 6 weeks before planting. Tuberous roots can be divided, but once settled the plants should not be transplanted. Poisonous plant. Prefers shade.

Montbretia (*Crocosmia* or *Tritonia crocosmaeflora*). Tender cormous plant. Lift corms and cormels and separate. Soak corms 48 hours before planting, if you store them over the winter.

Morning glory (*Ipomea purpurea*). Annual. Sow seeds where to grow in May, or earlier in pots indoors: for earlier bloom, not otherwise. Soak seeds in hot water 2 days before sowing and they will be germinating when you sow. (*I. batatus,* the tuber, is fun to grow in a glass of water, with the lower half in the water.)

Moss rose (*Portulaca grandiflora*). Annual. Plant outdoors after soil warms up; or in fall in warm climates. Sometimes self-sows. Resents transplanting.

Mugwort (*Artemisia vulgaris*). Perennial. Sow seeds in cold frame in late fall or sow outdoors anytime from spring to late summer. Division in fall easy. Stem cuttings.

Musk mallow (*Malva moschata*). Perennial. Rather weedy. *M. alcea* more desirable in the garden; divide or take stem cuttings.

Narcissus spp. Hardy bulbs. Lift in late summer after the leaves have died down. Divide offset bulblets. Replant. Also grown from seeds.

Nasturtium (*Tropaeolum majus*). Annual and perennial. Sow seeds outdoors. Resents transplanting. (If started in cold frame, use peat pots.) Germinates well in neutral soil at 68° to 70°F. Cuttings often root well. Start some for house plants. Used for pest-repellent.

Nerine lily—see Guernsey-lily

Nicotiana—see Flowering tobacco

Obedient plant—see False dragonhead; (See also Wild Flowers.)

Oconee-bells—see Wild Flowers

Orchid family (*Orchidaceae*). Perennials. Those with pseudobulbs: cut through rhizome between the pseudobulbs with sharp knife (or garden shears if very tough). Sterilize tools between cuts. Growing orchids from seeds very complicated and difficult. Consult specialty books. Do not try to adapt any to the garden except yellow lady's slippers (*Cypripedium pubescens* or *C. calceolus*). (See Yellow lady's slipper in Wild Flowers.)

Oriental poppy (*Papaver orientale*). Perennial. Sow tiny seeds, exposed to light; cover only slightly with sphagnum moss. Keep at 54°F. Root cuttings more satisfactory than division because taprooted poppies resent moving. Cut roots into 3- to 4-inch pieces, plant horizontally in the fall ½ inch deep in sandy loam. If division attempted, make deep, well-manured holes and plant with crowns 3 inches below surface. Grow annual poppies from seeds.

Pasque flower—see Anemone

Patience (*Impatiens balsamina, I. sultana*). Tender perennials grown as annuals. Also called garden balsam and sultan snapweed. Start tiny seeds indoors in February or March, uncovered but pressed into sphagnum moss. Take cuttings in late summer for winter pot plants. In mild climates, sow outdoors in April, and take cuttings during the summer months.

Peony (*Paeonia lactiflora*). Perennial. Divide, but only in fall, and plant in new ground, well-manured, with good gravel or other drainage material. Do not use peat or other acid fertilizers. Add lime if needed. Cut fleshy roots so each piece has several buds or eyes, and replant so that the top eye is *exactly* 1½ inches below surface of soil. Best to prepare hole 3 weeks in advance. If you plant peony seeds, they will take 1 to 2 years to germinate, with roots coming the first summer and shoots the second spring.

Peony, tree (*Peony suffruticosa*). Perennial. Easier to grow from seed than *P. lactiflora*. Stratify and plant in spring. These plants are difficult to grow. (See Ornamental Trees and Shrubs also.)

Periwinkle (*Vinca rosea*). Perennial. Propagation from stem cuttings easy. Layering. Division. Seeds grow indoors or outdoors if the temperature is at least 70°F.

Peruvian lily (*Alstromeria aurantiaca*). Tuberous plant. Division of tubers. Climate should be cool summers and mild winters.

Petunia (*Petunia* x *hybrida*). Annual. Pelleted seeds recommended. Or sow tiny seeds mixed with sand indoors in March in sterile starting mixture. Keep petunias pinched whenever leggy. Doubles and large-flowered varieties more difficult. Some self-sow. Stem cuttings in late summer for winter house plants.

Phlox (*Phlox paniculata*), perennial; (*P. drummondi*), annual. Try root cuttings for *P. paniculata*. Lift plant in fall, cut back larger roots to within several inches of crown; plant 2-inch pieces of root and set out late June of the following year. Divide crown or not and replant; water well. Stem cuttings in spring, July, or August. Keep suckers cut back; prevent self-sowing. Start *P. drummondi* seeds indoors in March. Or outdoors in May in open, sunny place, rather sandy soil. Can plant in fall if tiny seedlings protected over winter. In South plant in fall for winter annual. Save own seeds, stratify for 3 weeks at 40°F. before planting. Viable 1 to 2 years. For *Phlox divaricata* see Wild Flowers.

Pink—see Dianthus

Pitcher plant—see Wild Flowers

Plume-poppy (*Macleaya cordata*). Perennial. Also called bocconia. Spring-sown seed germinates best. Divide in spring. Root cuttings easy.

Poor man's orchid (*Schizanthus pinnatus*). Annual. Also called butterfly flower. Plant in greenhouse, or outdoors in May. Slow to germinate and to flower outdoors. Save own seeds, viable 4 to 10 years.

Poppy (*Papaver alpinum, P. glaucum*), the tulip poppy; (*P. macrostomum*), the bell poppy; (*P. nudicaule*), the Iceland poppy; (*P. rhoeas*), the Shirley poppy. Annuals. All grown from seed planted fall or spring. Many self-sow, especially Iceland poppy.

Poppy-mallow (*Callirhoe digitata*). Annual. Sow outdoors in May. Will do well in sunny, sandy, or poor soil. Divide spring or fall.

Portulaca grandiflora. Annual. Grow from seed in sunny place. Stem cuttings.

Pot-marigold (*Calendula officinalis*). Annual. Start indoors in March; or outdoors in April or May. Sometimes self-sows. Used for pest-repellent.

Potentilla (*Potentilla* spp.). Perennial. Some spp. also called cinquefoil. Gather seed in early fall to sow in pots. Softwood and hardwood cuttings in September and October. Divide in spring.

Prickly pear—see Wild Flowers

Primrose (*Primula veris*). Perennial. Also called cowslip primrose. Sow seeds in cold frame in fall, or in March place in cold frame for 3 to 4 weeks or in refrigerator deep-freeze compartment for 48 hours before sowing. Germination slow (4 to 6 weeks). Divide every 2 to 4 years, anytime after flowering, or in fall; remove offset rosettes to replant.

Rain-lily (*Cooperia* spp.). Tender bulbs. Lift and divide in fall. Replant in spring.

Red hot poker (*Kniphofia uvaria*). Perennial. Also called pokerplant and torch lily. Old plants can be divided; do not disturb young plants. Sow seeds in warm weather.

Rock cress (*Arabis* spp.). Perennial. Sow seed any time from spring to September. Softwood cuttings after flowering, in late spring. Divide spring or fall.

Rose—see Ornamental Trees and Shrubs

Rosybell—see Wild Flowers

Saint Bernard lily (*Anthericum liliago*). Stoloniferous perennial. Divide stolons. Done in North when lifted for the winter.

Salpiglossis (*Salpiglossis sinuata*). Annual. Also called painted tongue. Plant indoors in March. Do not cover seed; just press down into sphagnum moss. Water from below. Cover, and remove cover when sprouts come up. If sowing outdoors, wait until soil thoroughly warm. Sometimes self-sows. Try cuttings to winter over.

Salvia (*Salvia splendens*). Annual. Also called scarlet sage. Sow indoors in March with 70° to 85°F. heat. Prefers exposure to light to germinate. Outdoor sowing means 15 to 21 days for germination. Try cuttings. (See also Wild Flowers.)

Sand verbena (*Abronia fragrans*). Perennial. Also called pincushion flower. Self-layers where joints touch soil; divide. Other kinds of layering. Also crown division. Sow seeds indoors in April; outdoors in May. Expose seeds to light. Seedlings need night temperature of 55°F.

Saxifrage—see Wild Flowers

Scabious (*Scabiosa atropurpurea*), annual; (*S. caucasica*), perennial. Sow seeds of annual indoors in March with 70° to 85°F. warmth; outdoors when warm in May or later. Self-sows. Saved seeds viable 2 to 10 years. Sow seeds of perennial the same. Divide in spring.

Scarborough lily (*Vallota speciosa*). Bulbous plant. Tender; use in South or indoors. Divide bulblets when repotting, or after blooming in fall.

Scarlet sage—see Salvia

Scilla—see Squill

Sea lavender (*Limonium latifolium, L. carolinianum*), annuals; (*L. bonduelii*), perennial. Also called statice. Start annual seeds indoors in February or March, removing chaff; or outdoors in late spring. Long germinating and growing period needed. Also self-sows. Perennial seeds can be planted in cold frame in fall and put out early spring. (Some seeds have no embryo; are male.) Division. Root cuttings, which will provide 100 plants for 1.

Siberian wallflower (*Cheiranthus allioni* or *Erysimum asperum*). Biennial. In climates like that of England beautiful wallflowers, biennials or annuals, can be grown; but few places except Pacific Northwest are appropriate in United States. Sow in spring, and see whether you can make it grow.

Snakeroot—see Mistflower and also Black cohosh under Wild Flowers.

Snapdragon (*Antirrhinum majus*). Annual. Sow tiny seeds indoors in February or March. Do not cover seeds; just press down into sphagnum moss. Or sow in cold frame in early April, with germination in 14 to 21 days. Often self-sows. Saved seeds must be stratified at 40° to 42°F. in refrigerator for 3 weeks before planting.

Sneezewort (*Achillea ptarmica* 'The Pearl'). Perennial. This and other good cultivars 'Perry's White', 'Snowball', and 'Angel's Breath' easy to grow if sown in sterile conditions in spring, exposed to light for germination. Saved seeds viable for 3 to 4 years. Divide spring or fall.

Snow-in-summer (*Cerastium tomentosum*). Perennial. Sow outdoors early spring. Stem cuttings after flowering. Division, but you hardly need it; it is rampant.

Snow-on-the-mountain (*Euphorbia marginata*). Annual. Plant seeds outdoors, where they are to grow, as soon as ground workable. They self-sow, beyond your control probably.

Soapwort (*Saponaria calabrica*). Annual. Also known as Calabrian soapwort. Sow seed indoors in March and plant out in May. See also Bouncing bet.

Solomon's seal—see Wild Flowers

Spider flower (*Cleome spinosa*). Annual. Plant outside where it will grow. Just pat very small seed on soil. Protect for 10 to 21 days while germinating. Saved seeds must be prechilled at 40°F. in refrigerator for 48 hours. Self-sows. Try cuttings.

Spider lily (*Hymenocallis calathina*). Bulbous plant. Also called ismene and basket flower. Increase by offset bulblets when dug in fall for winter storage, or in South after flowering. Sometimes called *Lycoris radiata,* and yellow called *L. aurea. L. squamigera* called naked lady, because it blooms after foliage has died down.

Spiderwort, Virginia (*Tradescantia virginica*). Perennial. Sow seeds in spring. Self-sows. Divide in spring every few years. Stem cuttings in summer.

Spire lily (*Galtonia candicans* or *Hyacinthus candicans*). Tender bulbous plant. Also called summer hyacinth. Lift and divide bulblets in North, or if grown as a pot plant. In South, divide every few years. Easy to grow.

Squill (*Scilla* spp.). Hardy bulb plant. *S. nonscripta* is the English bluebell; *S. sibirica* is Siberian squill. Easy to propagate by removing offset bulblets (and they self-propagate freely). Lift and divide in fall. Sow seeds when ripe.

Star-of-Bethlehem (*Ornithogalum nutans*). Bulb plant. Spreads naturally; the problem is to control it.

Statice—see Sea lavender

Stock (*Mathiola incana*). Annual, or sometimes biennial. Sow seeds shallowly indoors or in cold frames in February. Protect. Saved seeds should be stratified for 3 weeks at 40°F. before planting. (Very slow to bloom so pelleted, trysomic 7-week stocks, available from some seedsmen, are recommended.)

Stoke's aster (*Stokesia laevis*). Perennial. Divide every 2 years. Stem cuttings. Sow seed indoors or out.

Stonecrop (*Sedum album* et al.). Perennial. Cuttings, division, and even broken pieces of leaf which will strike roots where in contact with soil. Sow spring or fall.

Straw-flower (*Helichrysum bracteatum*). Annual. Start seeds indoors or outdoors. Slow grower.

Summer cypress (*Kochia scoparia*). Annual. The instant shrub: if seed is sown outdoors in spring the plant will shoot up to 2 feet in a few weeks. Plant seed also in late fall. Self-sows.

Sunflower (*Helianthus annuus* et al.). Annuals. Plant seeds outdoors in spring; difficult to transplant. Birds and chipmunks plant it also for you if you have bird feeders. Try cuttings.

Sunrose—see Wild Flowers

Swamp lily or crinum (*Crinum amabile*). Large tender bulbous plants. Grown in South from bulblets divided from mother bulb in spring. Same for Jamaica crinum, hardy up to New York City and Long Island.

Swamp rosemallow—see Rosemallow under Wild Flowers

Swan River daisy (*Brachycome iberifolia*). Annual. Sow indoors in April; outdoors in late spring.

Sweet pea (*Lathyrus odoratus*), annual; (*L. latifolius*), perennial. Plant in October or very early in spring. Presoak annual seeds and sow in well-manured trench, as early as possible. Fill trench as plants grow until almost level. Put up wire support. Saved seeds need 3 weeks at 40°F. in refrigerator before sowing. Viable 2 to 8 years. For perennial, notch seeds. Divide in spring.

Sweet rocket (*Hesperis matronalis*). Biennial or perennial. Also called dame's rocket. Usually perennial, and self-sows. Better to rely on yearly planting in spring for blooms the following year.

Sweet sultan (*Centaurea imperialis*). Annual. Sow seeds directly in garden in May or when soil warms. In South can sow in fall. (See also Dusty miller.)

Sweet William (*Dianthus barbatus*). Biennial. Sow seed in February to set out the following year in garden. Can be carried over in deep flats or in nursery garden.

Tansy—see Wild Flowers

Tassel flower (*Emilia sigittata*). Annual. Also called Flora's paintbrush. Sow seeds outdoors where it is to grow.

Thrift (*Armeria maritima*). Perennial. Division very easy. Sow seeds in spring.

Tickseed—see Coreopsis

Tiger flower—see Mexican shell flower

Toadflax (*Linaria bipartita*), annual; (*Cymbalaria aequitrilobia*), perennial. Plant seeds of annual outdoors April or May. Make succession planting. Self-sows. Sow perennial seeds in spring. Cuttings in spring. Division in spring.

Torenia—see Blue wings

Transvaal daisy (*Gerbera* x *jamesonii hybrida*). Annual. Sow seeds in January indoors. Set out after frost gone. In South, treat as perennial.

Tree mallow (*Lavatera trimestris*). Annual. Plant in fall or in May after frosts where it is to grow. Do not transplant.

Tree peony—see Peony

Trout lily—see Faun lily

Tuberous begonia (*Begonia* x *tuberhybrida*). Treated as annual. Lift after blooming and gently twist off plant when dry. Store tubers in dry peat until March or April. Divide and plant with concave tops ½ inch below surface; transplant when 2 to 3 inches of top growth appear. Be sure each tuber has a bit of crown attached when you cut or twist it from the cluster.

Tulip (*Tulipa* spp.). Bulbous plant. Divide bulblets from mother bulb when lifted for summer storage after leaves die down. Replant in fall. Or lift and divide in fall. Protect with cages from rodents.

Turtlehead—see Wild Flowers

Twinspur (*Diascia barberae*). Annual. Plant seed indoors March; outdoors late April. Thin or transplant in May.

Valerian (*Centranthus macrosiphon*). annual; (*C. ruber*), perennial; (*Valeriana officinalis*), perennial. Sow *Centranthus* spp. indoors or outdoors in spring. Sow *V.*

officinalis in August and do not cover seed. Divide spring or fall from spreading rootstalk. *V. officinalis* also called garden heliotrope.

Verbena (*Abronia umbrellata*). Annual. Plant seeds indoors March with 80° to 90°F. heat. Or outdoors mid-May or when soil warm. Remove husks before sowing seeds. Saved seed is viable for 5 to 10 years. Take cuttings in summer. (See also Sand verbena.)

Veronica (*Veronica* spp.). Perennial. Also called spike speedwell. Divide spring or fall. Sow seed indoors in February; outdoors after frosts.

Vervain—see Wild Flowers

Violet (*Viola* spp.). Perennials and wild flowers. Division spring and fall. Stem cuttings mid-summer. Seeds of some species need exposure to light for germination. (See also Wild Flowers.)

Water-lily (*Nymphaea* spp.). Tuberous plant. Divide tubers in fall. Cuttings of stoloniferous growths at junction of stem and tuber. Plant in 4-inch pot. Seed used only for breeding; much space needed. (See also Wild Flowers.)

Windflower—see Wild Flowers

Winter aconite (*Eranthis hyemalis*). Tuberous plant. Divide tubers. Self-sows.

Wishbone flower—see Blue wings

Woodruff, blue (*Asperula azurea setosa*). Perennial used as ground cover. Division. Stem cuttings. Plant in moist soil in early May. Set in place when 6 inches tall.

Wood sorrel—see Wild Flowers

Yarrow (*Achillea millefolium*). Perennial. Division spring or fall. Sow seed in early spring exposed to light; outdoors later. Same for *A. filipendulina*.

Yucca filimentosa. Perennial. Separate suckers from mother plant. Root cuttings. Only one moth fertilizes this plant, so seeds are rare.

Zinnia (*Zinnia elegans*). Annual. Start seeds indoors March; plant outdoors April or May. Layer tall shoots in June: bend down when still pliant, fasten with wire loop like a hairpin, and the shoot will take root. To prevent spindliness of indoor seedlings, keep lights close to plants and reduce heat after germination.

Adder's tongue—see Trout lily; (See also Faun lily in Garden Flowers.)

Anemone, broad-leafed (*Anemone canadensis*). Also called meadow anemone. Sow seeds when ripe. Divide underground stems. Spreads. Needs woodsy habitat. See also Pasque flower, Rue-anemone, and Windflower.

Arrow-arum, Virginia (*Peltantra virginica*). Grow where wet. Squeeze out seeds and sow at once. Divide underground stems. Spreads.

Arrowhead (*Sagittaria latifolia*). Grow where wet, and divide the underground stems spring or fall. Spreads.

Aster spp. Save seeds in fall; sow in spring. Divide, moving part of clump and replacing the other part, unless very prolific. (See also Garden Flowers.)

Avens, water (*Geum rivale*). Sow seeds very early spring. Division, spring or summer. (Boil roots for chocolate-tasting drink.)

Baneberry (*Actaea alba, A. rubra*). Gather seeds in fall from berries; sow when ripe or sow in spring, with soil pH 6 to 7 for *A. alba*; soil pH 5 to 6 for *A. rubra*. Divide, keeping at least 2 eyes on each.

Barren strawberry (*Waldsteinia fragarioides*). Spreading ground cover plant with runners. Divide these; or collect seeds in fall and sow in spring. Prefers poor soil and sun.

Beach wormwood (*Artemisia stelleriana*). Also called dusty miller. Grow in sand, at beach. Pick stalks in October; keep dry in paper bag until time to sow seeds in spring.

Bearberry (*Arctostaphylos uva-ursi*). Evergreen ground cover for very special habitat: pH 4 to 5, sandy, and sloping. Push seeds out of berry in fall, stratify at 40°F. for 3

Wild Flowers

247

months; but germination erratic. Layering. Softwood and hardwood stem cuttings. Extremely difficult to transplant.

Bear's breech—see Garden Flowers

Bee balm or bergamot (*Monarda didyma, M. fistulosa*). Perennial. Divide; or sow spring or fall.

Bellflower—see Garden Flowers

Bishop's cap (*Mitella diphylla*). Also called miterwort. Sow seeds when ripe, in shady, moist place.

Bitterroot lewisia (*Lewisia rediviva*). This and other lewisias can be grown from seed or divided. Are on protected lists.

Black cohosh (*Cimicifuga racemosa*). Also called fairy candles, bugbane, snakeroot. Sow ½ inch deep in fall; winter over in cold frame. Then plant out; seedlings will bloom in 3 to 4 years. When established, divide so pieces have 2 eyes. Plant in somewhat acid soil.

Black-eyed Susan (*Rudbeckia hirta*). Annual or biennial which self-sows easily once established. Sow seeds in spring. Same applies to brown-eyed Susan (*R. triloba*) and the other rudbeckias.

Bleeding heart, plumy (*Dicentra eximia*). Sow seeds in August. Divide tuberous root when dormant. Needs some shade to grow well. Rare in the wild; available from nurseries.

Bloodroot (*Sanguinaria canadensis*). Rhizomatous plant. Sow seeds spring or fall; self-sows. Divide underground rhizomes.

Bluebell (*Campanula rotundifolia*). This campanula easy from seeds sown spring or fall. Divide underground stems. Spreads.

Blue cohosh (*Caulophyllum thalictroides*). Plant seeds in the fall or stratify in refrigerator for 4 weeks before planting. Prefers shade.

Bluet (*Houstonia caerulea*). Sow spring or fall, in slightly acid soil. Spreads in fields or unmown lawns in spring.

Boneset (*Eupatorium perfoliatum*). Sow in spring. Self-sows. Prefers moist places, but does wonderfully in rich garden loam.

Braun's holly fern—see Ferns

Bulblet bladder fern—see Ferns

Bunchberry (*Cornus canadensis*). Separate seed from pulp of berry when ripe. Sow at once ¾ inch deep.

Bunchflower (*Melanthium virginicum*). Also called bunch-lily. Sow seeds in fall. Only appropriate for very acid, boggy garden.

Burnet (*Sanguisorba canadensis*). Also called American burnet. Collect seed in fall, sow then or in spring in acid, wet, marshy place. Can divide in spring, keeping one piece, putting the rest back in its wild habitat. (Do not confuse with the salad herb, *S. officinalis*.)

Buttercup, swamp (*Ranunculus septentrionalis*). The native one; others escaped from introductions from Europe. Sow seeds in spring. If poor in germinating, divide. Needs acid, swampy meadow. Others include bulblet buttercup and mat buttercup, which is propagated by dividing creeping stems.

Butterfly weed—see Garden Flowers

Calla palustris. Also called wild calla. Sow seeds when ripe in wet place. Division. (Not the garden calla, which is *Zantedeschia aethiopica.*)

Campion, bladder (*Silene cucubalus*). Weedy but good for dried arrangements. You can't buy seed; collect and sow spring or fall; or just dig up a few plants if you see them.

Campion, Red (*Lychnis dioica*). Biennial. Sow seeds in April. Divide in spring. Same for *L. coronaria,* rose campion; *R. chalcedonica,* Maltese cross; and *L. flos-cuculi,* ragged robin. Cuttings also.

Canada lily (*Lilium canadense*). A worthy project would be to bring this back to wild damp meadows and lowlands. Plant seeds when ripe; protect for 5 years, especially from mice and other rodents. Divide scales from bulbs and plant in sandy, humus-rich soil until large enough to transplant.

Cardinal flower (*Lobelia cardinalis*). Plant seeds spring or fall. Or start inside in March. Short-lived, so take cuttings and save seed. Endangered; needs protection. Use mulch in winter if you live where it gets cold and soil heaves.

Chinese ladybell (*Adenophora lilifolia*). Also called lilyleaf ladybell. Propagate by cuttings or seeds. Resents transplanting.

Christmas fern—see Ferns

Cinnamon fern—see Ferns

Cinquefoil (*Potentilla canadensis*). Also called Oldfield cinquefoil. Divide runners. Sow seed while still moist, either soon after ripe, or preserve in moist peat moss.

Columbine (*Aquilegia canadensis, A. caerulea*). The first red, the second blue; the first Eastern; the second Western. Propagate them and help save them. Keep seeds moist and sow spring or fall in neutral woods soil. Division not so satisfactory as sowing seeds. (See also Garden Flowers and Wild Flowers sections.)

Common polypody—see Ferns

Coneflower—see Black-eyed Susan; see also Black-eyed Susan entries under Garden Flowers and Wild Flowers.

Coral-bells—see Garden Flowers

Corydalis (*C. aurea*), annual; (*C. bulbosa*), tuberous plant; (*C. flavula*), perennial, also called yellow fumewort. Seeds sown as soon as ripe; division if perennial or tuberous. Sometimes self-sows.

Crested shield fern—see Ferns

Daylily (*Hemerocallis fulva*). Escaped from cultivation, grows wild in many localities. More easily divided than raised from seeds, which take 2 years to come to flower. (See also Garden Flowers.)

Devil's bit (*Liatris pycnostachya*). Also called Kansas gayfeather. Division. Seeds in spring. *L. elegans* has corms.

Dogtooth violet—see Trout lily

Dragonhead, false—see False dragonhead in Garden Flowers section

Dutchman's breeches (*Dicentra cucullaria*). More likely to succeed from root cuttings and stem cuttings than from seeds. Needs protection.

Ebony spleenwort—see Ferns

Evening primrose (*Oenothera biennis*). Sow seeds in spring. (See also Garden Flowers.)

Evergreen woodfern—see Ferns

Fairy bells (*Disporum lanuginosum*). Sow seeds as soon as removed from pulp of the berry in fall. Needs fairly acid woodsy soil.

False aloe (*Agave virginica*). Gather seeds in fall; sow in spring.

False dragonhead—see Garden Flowers

False indigo (*Baptisia australia*). Also called wild blue indigo. Divide before or after flowering. Sow seeds in spring.

Foamflower (*Tiarella cordifolia*). Ground cover, perennial. Sow seeds when ripe in late summer, cool acid soil of pH 5 to 6. Divide in spring. Transplants successfully. Likes shade.

Forget-me-not—see Garden Flowers

Galax (*Galax aphylla*). Collect seeds in summer and sow. Divide creeping roots. Ground cover for acid soil, pH 5 to 6.

Gentian spp. Includes bottle gentian (*Gentiana andrewsi*), a perennial; the fringed gentian (*G. crinita*), a biennial; and the Rocky mountain fringed gentian (*G. thermalis*), an annual. Handle fine-as-dust seeds carefully. Sow as soon as ripe in pots to be put outside. They can be left outdoors all winter or brought indoors to a cool room. Transplant seedlings, if you are lucky enough to get any. Difficult.

Ginseng (*Panax quinquefolium*). Also difficult, especially to transplant. Grow from seed in sloping, semi-shady rather moist woods. Stem cuttings in summer. Root cuttings in spring.

Glade fern—see Ferns

Golden club (*Orontium aquaticum*). Warm-climate water plant. Collect seeds in later summer and plant in pot set in water for a while.

Golden pea (*Thermopsis carolinianum*). Sow seeds very early spring (not late spring) or fall. Self-sows when established. Divide, but note very deep roots.

Goldenrod (*Solidago* spp.). Division easy.

Goldie's fern—see Ferns

Ground nut (*Apios americana*). Plant tubers, which are good to eat, too. Sow seed in spring. Division of tubers spring. Long roots.

Groundsel (*Senecio aureus*). Also called golden groundsel and golden ragwort, a wetlands plant. Divide in late June after flowering. Stem cuttings. Root cuttings. Sow seed in summer. Self-sows. (See also Garden Flowers.) You can divide the rare *S. obovatus*, squaw weed, in spring.

Hay-scented fern—see Ferns

Indian cucumber root (*Medeola virginiana*). Separate seeds from pulp in September and sow at once. Moist, shady location.

Intermediate shield fern—see Ferns

Interrupted fern—see Ferns

Iris, crested (*Iris cristata*). Dwarf, likes part shade. Divide rhizomes. Same for *I. douglasiana; I. foliosa; I. fulva; I. giganticaerulea; I. pseudacorus,* the yellow flag; the beautiful Oregon iris (*I. tenax*); and other native, beardless irises.

Ironweed (*Veronia noveboracensis, V. altissima*). Plant seeds in spring where wet. Self-sows. Division in spring. Stem cuttings.

Jack-in-the-pulpit (*Arisaema triphyllum*). Sow seeds in fall, in acid (pH 5 to 6) soil and shady location, though it will survive in neutral soil. If your soil is somewhat dry, use green dragon (*A. dracontium*) instead. Do not handle the rash-causing bulbs. Do not eat, either.

Jacob's ladder (*Polemonium reptans, P. mellitum, P. carneum*). Plant seeds in fall in neutral soil. Often available from seedsmen or nurseries. (See also Garden Flowers.)

Joe-pye-weed (*Eupatorium maculatum, E. purpureum*). Sow spring or fall. Division. Needs moisture.

Lady fern—see Ferns

Lady's smock (*Cardamine pratensis*). Also called cuckoo flower and meadow bittercress. Sow seeds in summer, in wet boggy place. Division. *C. bulbosa* also.

Liverwort (*Hepatica americana*). Sow seeds spring or fall in damp, shady, acid soil (pH 5 to 6). Division of well-established clumps.

Long beech fern—see Ferns

Lungwort (*Pulmonaria officinalis*). Cowslip lungwort is *P. angustifolia*. Wild flowers in England and Europe, but useful in wild flower garden. Sow seeds late summer. Division.

Maidenhair fern; maidenhair spleenwort—see Ferns

Male fern—see Ferns

Marginal shield fern—see Ferns

Marsh fern—see Ferns

Marsh marigold (*Caltha palustris*). Sow in moist place. Spreads. Division. Stem cuttings.

Mayapple (*Podophyllum peltatum*). Gather seed and plant in fall. Separate seed from fruit by soaking and fermentation. Divide underground stems. This plant may need to be controlled.

Meadow-beauty (*Rhexia mariana, R. virginica*). Swamp flower; sow seeds in moist, acid soil. Divide in spring.

Meadow-rue (*Thalictrum dioicum, T. polygamum*). One short; one tall. Sow seeds as soon as ripe in fall or late summer in moist meadow soil. Divide every fourth year.

Meadowsweet (*Filipendula vulgaris*). An escapee. Divide in spring, or order seed from Canada or England. Same for *F. rubra,* called queen-of-the-prairie.

Merrybells (*Uvularia grandiflora* et al.). Sow seeds in fall. Divide early spring or late fall. Spreads in shady woods.

Monkeyflower (*Mimulus alatus; M. guttatus; M. ringens*). Wetland plants. Divide rootstalks yearly. Sow seeds late fall or early spring. Stem cuttings.

Monkshood (*Aconitum uncinatum*). Also called climbing or summer aconite. Sow seeds where soil is moist. Do not transplant. Poisonous. (See also Garden Flowers.)

Mountain spleenwort (*Asplenium montanum*). Do not grow; too fussy.

New York fern—see Ferns

Obedient plant (*Physostegia virginana*). See False dragonhead in Garden Flowers.

Oconee-bells (*Shortia galacifolia*). Sow seeds in fall; it thrives in rich, shady, moist soil. Divide often. Take cuttings in August. Makes a good ground cover, especially in southern climates. Mulch with oak leaves farther north.

Ostrich fern—see Ferns

Partridgeberry (*Mitchella repens*). Sow seed from berries in fall in acid (pH 4 to 5) soil. Cuttings in August. Needs shady woods.

Pasque flower (*Anemone pulsatilla*). Also called wild crocus. Sow seeds as soon as ripe in spring after soaking for 6 to 8 hours. Transplant to cool open woods or plains. Recommended pH 5 to 6.

Pea, bush or golden—see Aaron's rod in Garden Flowers

Penstemon—see Beard tongue in Garden Flowers

Phlox spp. wild. *Phlox divaricata* is woodland phlox; *P. sublata*, moss pink. Sow seeds in spring in neutral or only slightly acid soil. Difficult to collect; buy seeds. Take cuttings of side shoots after bloom of 1-year-old plants. Divide clumps in August; set 2 nodes deeper than before. Layer moss pink or take cuttings in August to put in moist sand.

Pitcher plant (*Sarracenia* spp.). Not recommended unless you have a true bog. Sow seed as soon as ripe in moist peat in pots kept in water until germination in 1 to 2 years. Grow in terrarium with sphagnum moss (and flies) until you can plant them in bog.

Pokeweed (*Phytolacca americana*). For fresh sprouts, gather seed from squashed (poisonous) berries and grow in box to bring in for the winter. Or plant in garden for spring vegetable.

Poppy-mallow—see Garden Flowers

Prickly pear (*Opuntia compressa*). Hardy cactus. Sow in dry sandy soil.

Primrose (*Primula mistassinica, P. parryi*). Sow seeds when ripe. Divide. See also Evening primrose.

Queen-of-the-prairie—see Meadowsweet

Rose, wild (*Rosa arkansana, R. blanda, R. californica, R. carolina, R. nitida, R. palustris, R. setigera, R. virginiana*). These native American roses are quite easily propagated by softwood and hardwood cuttings, and division; and species roses come true from seed. Sow as soon as ripe, cleaned from hips.

Rosemallow (*Hibiscus moscheutos*) and *H. palustris,* which is marsh flower. Root cuttings. Self-seeds. Sow in spring at waterside.

Rosette sage (*Salvia lyrata*). Sow seeds in fall or spring. Self-seeds freely. Needs sandy, woodsy, slightly acid soil.

Rosybell (*Streptopus roseus*). Cool woods and swamp edge plant. Clean seeds from berries and sow as soon as ripe. Fairly acid soil.

Rue-anemone (*Anemone thalictroides*). Sow in fall outside or in container indoors. Winter over before setting out in spring. Divide tuberous roots with 1 eye on each piece.

Salvia, blue and pink (*Salvia farinacea, S. rosea*). Also blue sage, which is *S. azurea*. Sow seeds indoors in February and move out after frosts. Division. Stem cuttings. (See also Garden Flowers.)

Sandwort (*Arenaria* spp.). Divide in early spring, especially the moss sandwort (*A. verna caespitosa*) used between flagstones. Seabeach sandwort (*A. peploides*) a bit fussy as to habitat, but once established self-seeds vigorously.

Saxifrage (*Saxifraga virginiensis*). Sow seeds in any warm season. Division in spring. *S. pensylvanica*: divide rhizomes. Beware of slugs.

Sensitive fern—see Ferns

Shooting star (*Dodocatheon meadia*). Sow seeds in rich woods soil in partial shade. Divide rosettes. Same for *D. amethystinum,* especially if you live in the Susquehanna valley in Pennsylvania. Or in the Mississippi valley.

Shortia—see Oconee-bells

Solomon's seal (*Polygonatum biflorum*). Whole or mashed berries sown in fall. Root division in fall.

Speedwell (*Veronica americana*). Also called brooklime speedwell. Many speedwells; this native prefers wet places for seeds sown in spring.

Spiderwort, Virginia—see Garden Flowers

Spinulose shield fern—see Ferns

Spring beauty (*Claytonia virginica*). Sow seed when fresh.

Division of small tubers; plant 2 to 3 inches deep. Self-sows.

Squirrel corn (*Dicentra canadensis*). Plant small tubers from roots 2 inches deep with wire mesh over the area (to fend off squirrels and chipmunks) and add mulch. (See also Dutchman's breeches.)

Star flower (*Trientalis borealis*). Sow seeds in fall when fresh in leaf mold. Leave out in flat over the winter. Divide tubers.

Stonecrop (*Sedum ternatum*). Sow in spring. Easy to grow. (See also Garden Flowers.)

Sunflower (*Helianthus* spp.). Easy from seed in spring. (See also Garden Flowers.)

Sunrose (*Helianthemum nummularium*). Grows only from seed. Plant where to grow.

Swamp pink (*Helonias bullata*). Bog plant. Sow in spring in very acid, moist soil. Easier to divide, once started. Needs protection.

Swamp rosemallow—see Rosemallow

Sweet woodruff (*Asperula odorata*). Increase by division.

Tansy (*Tanacetum vulgare*). Sow spring or fall. Self-sows. Division.

Tickseed—see Coreopsis in Garden Flowers

Trailing arbutus (*Epigaea repens*). Seed must be collected as soon as ripe before ants beat you to it. Rub seed capsules to get seeds out and plant in acid soil at once, or keep moist, sterile, and cool and plant in late fall in cold frame. Thin seedlings in summer. Leave for second winter. The following spring remove root ball intact or cut it into pieces and move to 3-inch pots. Water well for this summer, hold over another winter and plant out in its native habitat the following summer. Prevent drying at any time; prevent rot. And remember this requires such acid soil and such microorganisms that the hope of growing it anywhere besides its native habitat is slim. Much in need of protection. Stem cuttings also possible.

Trillium (*T. grandiflorum*). This is the only good one for gardens. Harvest seed from berries when ripe, keep moist, and sow outdoors in fall; or keep over to March and sow then in ½-inch leaf mold after 3 weeks in the refrigerator. Needs protection.

Trout lily (*Erythronium* spp.). Collect seeds and keep over winter in refrigerator; or plant fresh in leaf mold. Germination slow and poor. Plant small bulbs, or bulblets, if you can find them, in late summer or fall. (See also Faun lily in Garden Flowers.)

Turtlehead (*Chelone glabra*). Collect seed in October and sow in spring. Once established, will self-sow in wet, cool place.

Vervain (*Verbena hastata, V. canadensis*). Sow seeds spring or fall. Layering; trailing branches do it naturally. Seeds slow, but plants self-sow. Division of young plants. Stem cuttings in July. Use for ground cover even where soil poor.

Violet (*Viola* spp.). Sow in slightly acid soil. Collect seeds from cleistogamous bulgings, or flowers, near base of plant. These are unopened and greenish. Watch out; they twist open before you know it. Pick whole pods into paper bag, so they will pop open inside. Self-sows. Division less satisfactory. (See also Garden Flowers.)

Virginia bluebells (*Mertensia virginica*). Sow seeds spring or fall. Self-sows.

Water-lily (*Nymphaea* spp.). For breeding new varieties, hand-pollinate and sow resulting seed in moist soil. (Harvest pollen from staminate plant the day before mother flower opens and is receptive. Keep both flowers enclosed in plastic bags until time of actual pollination.) For usual propagation, divide underwater creeping stems. (See also Garden Flowers.)

Wild geranium (*Geranium maculatum*). Also called spotted cranesbill. Sow spring or fall. Divide spring or fall, leaving 3 to 4 eyes on each piece.

Wild hyacinth (*Camassia scilloides*). Also called Atlantic camas. Sow seeds in fall. Lift bulbs when overcrowded in late fall, divide and replant.

Wild lily-of-the-valley (*Maianthemum canadense*). Also called Canada mayflower. Plant berries in fall in woodsy soil. Divide when clumps grow very thick.

Wild senna (*Cassia marilandica*). Sow only fresh-picked seed. Division.

Willow amsonia (*Amsonia tabernaemontana*). Gather seeds in fall and sow in spring. Self-sows.

Windflower (*Anemone quinquefolia*). Sow where soil is cool and moist in spring. Divide in spring.

Wood sorrel (*Oxalis montana*). Gather seed in fall. Sow in spring in cool, woodsy place, where soil is pH 4 to 5. Divide clumps or bulbs.

Yarrow—see Garden Flowers

Yellow bead lily (*Clintonia borealis*). Collect seed in summer and keep cool until spring. Sow in spring in cool woods, where soil is acid (pH 5 to 6). Divide rhizomes in spring.

Yellow bedstraw (*Galium verum*). Collect seed of this dye and rennet plant in summer and sow the following spring. Self-sows.

Yellow flag—see Iris, crested

Yellow lady's slipper (*Cypripedium pubescens* or *C. calceolus*). These are the only native orchids I recommend for the average home gardener. They are the only ones that can succeed in most gardens. Other wild orchids cannot live unless conditions and underground microorganisms present are very, very special and exactly appropriate. Yellow lady's slippers are less fussy. Place seeds in early winter in bulb pans. Put in cold frame in acid woodsy soil with pH 6 to 7. When planting, ease contents of bulb pan into outdoor spot where soil is near to flower's native habitat. But sometimes garden soil, in partial shade, with leaves decaying on it, will do. Division, with equal care.

Zephyr-lily (*Zephyranthes atamasco*). Tender bulb. Sow seeds in South in April. Separate bulbs in early summer.

For detailed information on propagating wild flowers from seed, see chapter 6; for propagation of wild flowers by layering, see chapter 3.

Blackstemmed spleenwort (*Asplenium resiliens*). Not recommended because it requires very special habitat in a rock crevice.

Boston fern—see House Plants

Bracken or brake (*Pteridium aquilinum*). Nobody really wants to grow this fern unless poor, weedy land needs greening. Rhizome division easy. Spreads quickly.

Braun's holly fern (*Polystichum braunii*). Excellent fern to grow in cool shady place. Mist-spray if hot. Crown division in fall. Spores. Do not collect, unless you plan to propagate and put back, because it is rare.

Broad beech fern (*Thelypteris hexagonoptera*). This fern, for the South, and the related long beech fern (*T. phegopteris*), for the North, are easy to grow and propagate. Rhizome division and propagation by spores.

Bulblet bladder fern (*Cystopteris bulbifera*). Also called berry fern. If you have limestone rocks, grow and propagate this fine fern by removing small bulblets from base of blades and planting them in the rock crevices. Or you can just let them fall to the ground and transplant them when they have 4 leaves.

Chain fern (*Woodwardia* spp.). Good for wet places, especially if brackish. Divide rhizomes; easier with thinner nets on netted chain fern (*W. areolata*) than with big tough growths on Virginia chain fern (*W. virginica*). Spores.

Christmas fern (*Polystichum acrostichoides*). Rootstalk division. Spores.

Cinnamon fern (*Osmunda cinnamonea*). Careful division of crowns or creeping rootstalks. Spores. Put in damp, acid soil, in shade or sun.

Ferns
(See also Ferns under House Plants.)

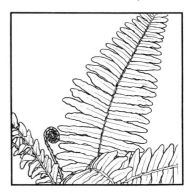

Common polypody (*Polypodium virginiana* or *P. vulgare*). Small fern for rocky place; divide creeping rootstalk. Spores.

Crested shield fern (*Dryopteris cristata*). Also called crested fern. Divide creeping scaly rootstalk. Spores.

Ebony spleenwort (*Asplenium platyneuron*). The only recommended spleenwort. Divide rhizomes. Spores. Prefers limestone or sandstone nearby.

Equisetum—see Horsetail

Evergreen woodfern—see Marginal shield fern

Glade fern (*Athyrium pycnocarpon*). Also called silvery glade fern or narrow-leaved spleenwort. Divide creeping rootstalk. Spores.

Goldie's fern (*Dryopteris goldiana*). Divide rootstalk of this big woods fern. Spores. Keep always moist. Propagate and put back. Rare.

Hay-scented fern (*Dennstaedtia punctilobula*). Rhizome division. Spreads rapidly if snails and thrips stay away. Spores.

Horsetail (*Equisetum* spp.). Annual aboveground; perennial below. Divide rhizomes. Spores from fertile stems. (These are the plants often ground up to use as sprays, they have lots of silica. See Cleaning and Sterilizing Equipment in chapter 1 for instructions on making an *Equisetum* solution.)

Intermediate shield fern (*Dryopteris intermedia*). Crown division. Spores.

Interrupted fern (*Osmunda claytonia*). Large where soil acid; small where it's not. "Interruptions" on stalk are where fertile leaflets grow. Spores. Divide very tough matted rootstalk.

Lady fern (*Athyrium filix-femina*). Divide rootstalk. Spores.

Long beech fern—see Broad beech fern

Maidenhair fern (*Adiantum pedatum*). Divide creeping rhizomes. Spores.

Maidenhair spleenwort (*Asplenium trichomanes*). Not recommended because it requires special conditions in crevices in limestone rocks.

Male fern (*Dryopteris filix-mas*). Divide rhizomes. Spores. Needs moisture and stony place. Do not collect, unless you plan to propagate and put back.

Marginal shield fern (*Dryopteris marginalis*). Also called evergreen woodfern, marginal woodfern, and leather woodfern. Crown division in spring. Spores.

Marsh fern (*Thelypteris palustris*). Also called snuffbox fern. Divide creeping rootstalk of this wetlands, common fern. Spores. Spreads rapidly where swampy and acid.

New York fern (*Thelypteris nove-boracensis*). Divide creeping rootstalk. Spores.

Ostrich fern (*Matteuccia pensylvanica* or *M. struthiopteris*). Divide this large, streamside or moist-land fern by crown division. Spores.

Sensitive fern (*Onoclea sensibilis*). Also called bead fern. Divide this ditch-side, rampant fern by cutting creeping rootstalk. Spreads fiercely in somewhat acid soil, whether in sun or shade.

Spinulose shield fern (*Dryopteris spinulosa*). Also called spinulose woodfern. Woodland swamp fern propagated by dividing creeping rhizomes. Spores.

Woodsia spp. Rusty woodsia (*W. ilvensis*) is also called resurrection fern for it wilts down in dry weather and rises again when it rains. Needs acid soil. Blunt-lobed woodsia (*W. obtusa*) needs neutral soil. Others need limestone, high-altitude habitat. Divide rhizomes. Spores.

Vegetables

Most vegetables are easily grown from seed. Start them directly in the garden in the spring, or earlier in flats indoors to get a head start on an otherwise short growing season. Transplant the seedlings when they are vigorous and the weather is warm enough for them. (See hints in chapter 6.) Consult seed catalogs and choose the varieties most suited to your own area and needs. Vegetables have been carefully bred and the seeds and plants available from seed companies and nurseries are reliable products of years of selection.

Learn which vegetables can be planted early, which need to be planted late. For example, lettuce, peas, broccoli, onions, spinach, and turnips can go in 6 weeks before the frost-free date for your area. Beets, carrots, radishes, chard, parsnips, and mustard greens, 2 to 4 weeks earlier. But wait at least until a week after the frost-free date to plant lima beans, eggplant, peppers, sweet potatoes, and tomatoes.

It's best to start with purchased seeds and plants, especially if you need specific days-to-maturity and disease-resistant qualities. But try also collecting seeds from your own plants, for this kind of experimenting may be fun and very rewarding. (See chapter 6.) Collect seeds for this purpose from nonhybrid plants, but even so you may create hybrids if you do not control cross-pollination between related plants such as the *Brassicas*, *Cucumises*, and *Cucurbits*.

All vegetables listed are annuals, unless otherwise noted. Do not ignore the perennial vegetables; once established they will produce for many years and are easily multiplied by root division.

Artichoke, globe (*Cynara scolymus*). Perennial. Grown mostly in warm climates where winter temperatures are above freezing, though roots are hardy along the Atlantic seacoast up to Massachusetts. Propagated by rooted suckers or root division.

Artichoke, Jerusalem (*Helianthus tuberosus*). Perennial. Not an artichoke, but a tuber. Plant tubers spring or fall in sunny or semi-shaded fertile location. The next fall tubers will have multiplied; dig as needed. Those remaining will send up shoots in spring. Spreads rapidly, hard to get rid of.

Asparagus (*Asparagus officinalis*). Perennial. Fully 3 years required for harvestable asparagus but it's well worth the wait. Plant year-old roots 10 inches deep in well-manured trenches in early spring. Or start from seed. Try seeds from own most productive (nonhybrid) plants. In late summer, collect and squash berries, let stand in water a few days, clean and dry seeds. Store dry. Highest germination ratio at 85°F. Start in flats and transplant when seedlings vigorous.

Beans, bush (*Phaseolus nanus*) **and climbing** (*P. vulgaris*). Plant seeds, preferably inoculated with nitrogen-fixing bacteria, in spring when soil warm. Germination 7 to 10 days at 65°F. Beans thrive in warm weather. As they self-pollinate, named varieties are likely to produce seeds true to type. Let them dry on the plant. Store dry. Seeds viable 4 years.

Beets (*Beta vulgaris*). Biennial. Plant the frost-hardy seeds very early in spring directly in garden, or plant later in summer in time for fall maturing. They mature in 40 to 65 days; like cool weather. To save seeds, store only good beets in cool cellar; plant again in spring (prevent cross-pollination with chard) and allow to go to seed. Viable 6 years.

Broccoli (*Brassica oleracea botrytis cymosa*). Cool weather plant. Start from seed indoors 6 weeks before last frost for spring crop. Where summers are hot, better started mid-summer for fall crops. Matures in 60 days. Broccoli will go to seed if left to flower but crosses easily with all other *Brassicas*. Seeds viable 5 years.

Brussels sprouts (*Brassica oleracea bullata gemmifera*). Biennial. Need cool fall weather for good sprouts; will last after frost. Start seeds indoors in spring 8 weeks before last frost date. Germination 6 to 10 days at 50 to 60°F.

Cabbage (*Brassica oleracea capita*). Biennial. Cool weather crop. Start seeds indoors 4 to 6 weeks before last frost date for spring crop. For fall crop sow seeds mid-summer in garden or cool greenhouse. Germination 6 to 10 days at 60°F. To save seeds, cover good plants with thick mulch to protect through winter; will bloom the next summer and go to seed. But cross-pollination with other *Brassicas* is likely if any in bloom.

Cabbage, Chinese (*Brassica pelinensis*). If springs are long and cool, start from seeds indoors 4 weeks before last frost date. Difficult to transplant; disturb roots as little as possible. Will bolt and go to seed in hot weather. For fall crops, start late summer in garden where to grow. Germination 4 to 5 days; matures in 70 to 80 days. Plants will stand a light frost.

Carrot (*Daucus carota sativa*). Biennial. Sow seeds ½ inch deep in well-cultivated sandy loam in early spring. Do not transplant. Germination 2 to 3 weeks. Matures in 65 to 70 days. For seeds, mulch good plants heavily to protect through winter and allow to go to seed next summer. Beware of cross-pollination with wild carrot (Queen Anne's lace). Seeds viable 4 years.

Cauliflower (*Brassica oleracea botrytis*). Treat this vegetable same as broccoli.

Celery and celeriac (*Apium graveolens*). Start seeds indoors 8 to 10 weeks before last frost date. Germination 8 to 12 days at 60°F. Soak seeds for 24 hours for better germination. Young seedlings are very delicate. Transplant in spring to fertile, highly organic soil. Matures in 100 to 120 days. In South grow during winter.

Chard, Swiss (*Beta vulgaris cicla*). Biennial. Sow seeds 1 inch deep in garden soon after last frost. Germination 8 to 10 days. May soak seeds to hasten sprouting. If mulched, chard will last into winter. Seeds viable 6 years. Also try ruby chard.

Collards (*Brassica oleracea acephala*). Biennial. More heat-tolerant than other *Brassicas*. Culture otherwise same as for cabbage. In South may be grown most of the winter.

Corn, sweet (*Zea mays saccharata*). Plant seeds 1½ inches deep in garden when soil has warmed in late spring. Germination 5 to 8 days. Corn is wind- and water-pollinated and plots should be in squares at least 3 rows deep to insure complete pollination. Matures within 80 to 95 days. Let some ears (only nonhybrid) ripen fully if breeding seed is wanted. When several varieties of corn are grown together, cross-pollination and poor seed result unless plants are kept a good distance apart. Seeds viable 2 years.

Cucumber (*Cucumis sativus*). Plants seeds 1 inch deep after soil is warm in late spring. Seedlings do not transplant well; if started indoors use individual peat pots. Begins to bear fruits in 55 to 65 days. If seeds are wanted, let some early well-formed cukes ripen on vine, remove, clean, and dry seeds. But beware: will cross easily with melons and other cucumbers. Seeds viable 10 years.

Eggplant (*Solanum melongena*). Most successful in warm climates. In temperate areas start seeds indoors 8 to 10 weeks before last frost date. Soak seeds 24 hours before planting. Germination 10 to 12 days. Matures in 120 days. Seeds viable 6 years.

Endive (*Chicorium endivia*). Germination 8 to 10 days. Culture same as for lettuce. Seeds viable 10 years.

Garlic—see Herbs

Kale (*Brassica oleracea acephala*). Biennial. Likes cool weather; may be grown through winter as far north as Pennsylvania. Start sowing seeds in garden as soon as soil is workable; make several plantings. Germination 8 to 10 days at 60°F. Matures in 55 to 65 days.

Kohlrabi (*Brassica caulorapa*). Biennial. Start seeds early in spring or later in mid-summer for fall harvest. Likes cool weather. Germination 7 to 10 days. Matures in 85 days. In South may be grown through the winter. Seeds viable 5 years.

Leek (*Allium porrum*). Sow seed ½ inch deep in garden early spring. Germination 8 to 10 days. Matures in 130 days. Seeds viable 3 years.

Lettuce, head and loose-leaf (*Lactuca sativa*). Sow seeds as soon as ground workable; they tolerate a light frost. Germination 6 to 8 days at 60°F. Seeds need light to germinate. Plant throughout spring and again in fall. Requires rich fertile soil and cool weather. Head lettuce matures within 80 days. Leaf varieties mature in 50 days and they tolerate higher temperatures. If seeds are wanted, let good specimens bolt during summer; collect seeds in fall. Seeds viable 5 years.

Muskmelon (*Cucumis melo cantalupensis*). Plant seeds 1½ inches deep in hills in late spring when soil warms or start indoors in peat pots. Transplanting difficult. Needs long dry season of 80 to 100 days to mature. Save and dry seeds from mature good fruit if you like, though crossing with other *Cucumis* plants likely. Seeds viable 6 years.

Mustard greens (*Brassica juncea*). Plant seeds anytime after soil is warm. Germination 6 to 8 days. Matures in 30 days; seeds viable 6 years.

Okra (*Hibiscus esculentus*). Warm weather plant. Sow seeds where to grow in late spring. Germination 6 to 8 days. Matures in 65 days. For seeds, let pods fully ripen.

Onion (*Allium cepa*). Takes 14 to 15 warm weeks from seeds to good-sized dry onions. If season long enough, start seeds in garden early spring. May also be started indoors in flats, then treated as onion sets. Sets, which are often purchased, are planted 1 inch deep in good, neutral garden soil. For scallions, sow onion seeds thickly in garden, pull in 8 to 9 weeks. Seeds viable 2 years. You can grow your own onion sets by planting onion seeds late, crowded very near together and in poor soil. Break down tops, as you do for large onions to promote drying off for harvest. Try also multiplier and Egyptian onions. The first kind will propagate itself into several new stems and little bulblets. Egyptian or "top" onions have bulbils instead of flowers at the tip of their big hollow stems. When mature, the weight of the little cluster bends over the stem to the ground where the bulbils then plant themselves; or you can separate the small bulbs when young and plant them yourself. Shallots are a variety of onion, though formerly called *Allium ascalonicum*. They produce large clusters of bulblets, which you divide and replant in order to propagate shallots.

Orach (*Atriplex hortensis*). A leafy green, yellow, or red annual vegetable, very tall. Tolerant of alkaline soils. Propagate by seeds. Ask for *A. cupreata* if you want red leaves. Very decorative. Goes to seed rapidly; replant biweekly.

Parsnips (*Pastinaca sativa*). Biennial. Germination 2 to 3 weeks. Plant seeds very early spring. Matures in 100 days. Leave roots in ground well into fall or over the winter; winter frost seems to improve texture and flavor, but eat in spring before green growth starts. Inedible after that. Seeds viable 3 years.

Peas (*Pisum sativum*). Cool weather crop; high temperatures inhibit germination, growth, and podding. Plant seeds as soon as garden can be worked in spring. Good idea to inoculate seeds with a dried preparation of nitrogen-fixing bacteria unless peas have been grown in the same spot the year before. Another crop may be started in late summer if your fall is long. Matures in 60 to 80 days. Peas self-pollinate and some varieties produce seeds worth saving. Seeds viable 3 years.

Peppers, sweet and hot (*Capsicum* spp.). Best results in warm climates. Need 3 months sunny weather. In North start seeds indoors 6 to 8 weeks before last frost date. Transplant very carefully. Seeds may be saved from ripened pods. Seeds viable 4 years.

Potato (*Solanum tuberosum*). Cut seed potatoes in 2 to 4 pieces, each with at least 1 eye; or plant whole. Plant 4 inches deep in fertile garden soil, or under hay mulch as Ruth Stout does. Plants like cool growing season; start very early in spring if summers hot. Tubers have reached maximum growth when vines die down; must harvest before frost. Save good tubers for next year's seed.

Pumpkin (*Cucurbita pepo*). Plant seeds in hills 2 to 4 weeks after last frost. Germination 8 to 10 days. Likes long warm summer. Matures in 110 to 120 days. Save seeds of nonhybrid varieties if desired but cross-pollination with squashes or other *Cucurbits* likely. Seeds viable 5 years.

Radish (*Raphanus sativus*). Matures in 3 to 6 weeks in cool weather. Seeds very hardy, start in early spring. Germination 4 to 6 days. Thin vigorously. Save seeds from late-bolting radishes. Seeds viable 5 years.

Rhubarb (*Rheum rhaponticum*). Perennial. Start perennial bed with dormant crowns in early spring. Divide every 5 years to maintain vigorous plants.

Soy-beans (*Glycine max*). Plant seeds 2 to 3 weeks after last frost 1 inch deep in garden. Inoculate seeds with nitrogen-fixing bacteria preparation if soy-beans have not been planted in same place before. Will mature in 95 to 115 days. Save vine-ripened seeds of strong plants. Seeds viable 2 years.

Spinach (*Spinacia oleracea*). Grows best in cool weather. Plant seeds ½ inch deep in garden in early spring. Germination 5 to 9 days. Matures in 40 to 60 days. Fall crop may be planted also. Seeds viable 5 years.

Squash, bush and trailing (*Cucurbita* spp.). Plant seeds in garden after soil well warmed; roots resent disturbance. Germination 8 to 10 days. May collect and save seeds of ripened fruit but crosses with other squashes and pumpkins likely. Seeds viable 4 years.

Sweet potato (*Ipomoea batatas*). Needs 4 months warm weather to mature; grown mostly in warm climates. "Sprout" untreated whole potatoes indoors. Cover with 2-inches sand and keep moist. When sprouts are 4 to 5 inches tall, pull and plant in garden.

Tomatoes (*Lycopersicon esculentum*). Plant seeds ½ inch deep in flats indoors for early tomatoes and directly in the garden for more vigorous but later fruit. Germination 8 to 10 days. Matures in 80 to 100 days; likes warm sunny weather. Save seeds from nonhybrid varieties. Seeds viable 4 years. Cuttings easy.

Turnips (*Brassica rapa*). Biennial. Need cool weather for good roots. Start seeds early in spring where plants are to grow. Germination 4 to 7 days. Start fall crop in mid-summer and harvest after first light frost, or in late fall. Seeds viable 5 years.

Herbs

Since many herbs are of Mediterranean or Middle Eastern origin, they thrive in full sun and in soil that is not too rich. But there are exceptions, indicated in the list below. Some herbs manage to do very well in rich, organic soil, with very little lessening of the essential oils, though such loss is said to be one adverse effect of growing herbs where the soil is rich.

Herbs have not been bred and hybridized as vegetables and many ornamentals have; most are quite the same plants they have been for centuries. Propagation by seed, therefore, is often a simple matter of collecting and planting; and many self-seed freely. Perennial herbs are easy to divide; and some spread so rapidly that they are practically a menace. Do not shun the wild or escaped herbs for they, too, will add to the charm and interest of your garden. If you know where they grow and it is legal to take them, collect seeds or a division of the plant, and include them also. Be lavish with herbs. Use many that attract bees. Grow them not only in herb gardens, but also as companion plants to protect your vegetables and delicate ornamentals, in the border, under trees, in the rock garden, and wild garden. They are truly multipurpose, and all gardeners, I believe, can be thankful that they are so easy to propagate.

Angelica (*Angelica archangelica*). Biennial, if left to bloom. Plant seeds after mature in August in a seedbed or flat. Do not cover seeds. Germination in 21 to 28 days at 70°F. Move to rich soil. Seeds viable only 1 year.

Anise (*Pimpinella anisum*). Annual. Sow seeds directly in garden in late spring. Do not transplant. Cover seed with ¼ inch soil. Germination 5 to 6 days, 70°F. Attracts bees.

Basil (*Ocimum basilicum*). Annual. Very frost sensitive. Start outdoors in late spring but keep seedlings shaded. May be started in flats earlier indoors. Germination 7 to 10 days. Needs rich soil. Keep tops pinched for bushy growth. Take cuttings in fall for house plants.

Bay (*Laurus nobilis*). Keep this tree as potted plant outdoors in summer and indoors in winter. Seeds germinate within 4 weeks at 75°F. Stem cuttings easily rooted, but may take up to 6 months.

Bergamot or bee balm (*Monarda didyma*). Perennial. Seeds difficult; seedling growth very slow. Best started from and increased by root division. Look for wild plants to bring into the garden. Will spread rapidly; divide every 3 years. Likes rich soil. Attracts bees.

Boneset *(Eupatorium perfoliatum).* Perennial. Collect wild plants in the fall and set in the garden. They will flourish and seed themselves. Divide root clumps in the fall for more plants if wanted. Will grow large in rich soil.

Borage (*Borago officinalis*). Annual. Sow where to grow; it will later self-seed. Attracts bees.

Burnet (*Sanguisorba minor*). Perennial. Will grow easily from seed and self-sows once established.

Calendula or pot-marigold (*Calendula officinalis*). Annual. Sow seeds in April or May when soil is at least 60°F. Thin to 10 inches apart.

Caraway (*Carum carvi*). Biennial. Does not tolerate transplanting; sow directly in the garden in early spring. Seeds ripen the following fall.

Catnip (*Nepeta cataria*). Perennial. Sow seeds in garden early spring or fall. Easily propagated also by root division and stem cuttings. Requires rich moist soil. Attracts bees.

Chamomile, German (*Matricaria chamomilla*). Annual. Broadcast seeds in the garden in the fall; they will germinate the following spring. Attracts bees.

Chamomile, Roman (*Anthemis nobilis*). Perennial. Start from seeds or from root divisions. It will spread by runners, especially if not allowed to flower. This low, creeping chamomile is sometimes used for lawns. Attracts bees.

Chervil (*Canthuiscus cerefolium*). Annual. Sow seeds directly in the garden in early spring. Resents transplanting; prefers moist soil and partial shade.

Chives (*Allium schoenoprasum*). Perennial. Start seeds outdoors in spring or indoors in flats before last frost. Cover seeds with a sprinkling of fine soil or ground sphagnum. Germination 10 to 14 days. Dense clumps of mature chives can be divided in the spring. Self-sows.

Comfrey (*Symphytum officinale*). Perennial. Start from root cuttings in spring or fall. Likes rich, limed soil. It will spread and established plants can be divided in the spring.

Coriander (*Coriandrum sativum*). Annual. Sow seeds in spring where plants are to grow. Matures 90 to 100 days. Watch maturing seeds as they are easily lost.

Costmary (*Chrysanthemum balsamita*). Perennial. Produces no seed and must be propagated by division. As with most chrysanthemums, this is easily done in the spring.

Dill (*Anethum graveolens*). Annual. Sow seeds directly in the garden; does not tolerate transplanting unless very young. Germination 7 to 10 days. Matures in 70 days. Will often self-sow in mild climates.

Fennel (*Foeniculum vulgare*). Biennial. Start from seed in early spring. Germination in 2 weeks; thin to 6 inches apart. Will self-sow if allowed to go to seed. Incompatible with some vegetables; isolate.

Garlic (*Allium sativum*). Perennial bulbs. Break bulbs into individual cloves; plant base down 1½ to 2 inches deep in well-worked soil. Start in the fall for next fall's harvest or early in the spring. The flower head produces bulbils which you can also plant.

Geranium, scented (*Pelargonium* spp.). Perennial. Best started from stem cutings in the fall and then set in the garden the following spring. Plants may also be brought inside for the winter.

Germander (*Teucrium* spp.). Perennial. Best propagated by cuttings taken early in growing season.

Horehound (*Marrubium vulgare*). Perennial. Start in garden in the spring from seeds or earlier in the cold frame. Cuttings and root divisions easy. Thrives in poor soil and in sun.

Horseradish (*Armoracia lapathifolia*). Perennial. Produces no seeds but very easily started from the smallest root cutting. Prefers damp clayey soil; will spread.

Hyssop (*Hyssopus officinalis*). Perennial. Start indoors from seeds. Germination (in dark) in 10 days at 70°F. Cuttings. Division. Attracts bees.

Lavender (*Lavandula* spp.). Perennial. Start from seed indoors in February; transplant seedlings when they are strong and weather is warm. Germination takes 2 weeks. Cuttings made in August will be ready for the garden the following spring.

Lavender cotton (*Santolina chamaecyparissus*). Perennial. Start from seeds in spring. Cuttings from new growth may be rooted. Not reliably hardy north of New York City.

Lemon balm (*Melissa officinalis*). Perennial. Easily started from seed and will reseed itself. Spreading by runners rapid; plantings should be occasionally divided. Cuttings easily rooted. Attracts bees.

Lovage (*Levisticum officinale*). Perennial. Start seeds late summer when ripe or following spring. Divide mature plants in very early spring to maintain their vigor.

Marjoram (*Majorana hortensis*). Tender perennial. Treat as annual. Start seeds indoors in flats, transplant when soil has warmed in spring to garden. Attracts bees.

Mints (*Mentha* spp.). Perennial. May be started from seeds which do not come true. Cuttings more reliable. Plants will spread and should be divided every few years. All mints prefer moist rich soil and a semi-shaded location.

Nasturtium (*Tropaeolum* spp.). Perennial, but treat as annual. Sow seeds in late spring. Cover with 1 inch of soil. May reseed themselves. Used for pest-repellent.

Oregano (*Origanum vulgare*). Perennial. Is a wild marjoram except oregano is much hardier. May also be started from cuttings. Attracts bees.

Parsley (*Petroselinum crispum*). Biennial. Soak seeds 24 hours, then plant where to grow. Slow to germinate. Likes moist soil, partial shade. Can be potted in fall, cut back, and new leaves will grow. Be careful of taproot.

Rosemary (*Rosmarinus officinalis*). Tender perennial. Seeds very difficult; best results from stem cuttings taken in August. Pot up and water faithfully. In North, move indoors for the winter.

Rue (*Ruta graveolens*). Perennial. Easily propagated from seeds started indoors in February. Also may be started from cuttings taken in the spring.

Safflower (*Carthamus tinctorius*). Annual. Sow seeds in April where they are to grow. Do not transplant. Needs 4 months' sun.

Saffron (*Crocus sativus*). Perennial bulb. Plant bulbs late July or August. Divide every 3 years.

Sage (*Salvia officinalis*). Perennial. Large seeds are easily started indoors in flats in March or in garden in April. Germination 15 to 20 days. Stem cuttings in early fall.

Summer savory (*Satureja hortensis*). Annual. Broadcast seeds in garden in spring; light needed for germination (7 to 10 days). Thin seedlings to 1 foot apart.

Sweet cicely (*Myrrhis odorata*). Perennial. Seed may be sown in fall, though germination rate is poor. Once established it may self-seed. Divide mature plants in spring. Grow in a moist, shaded location.

Sweet woodruff (*Asperula odorata*). Perennial. Difficult to grow from seed. Purchased plants will soon spread and may be divided. Requires rich moist soil and shade.

Tansy (*Tanacetum vulgare*). Perennial. Difficult to start from seed, but once established, will spread rapidly. Division easy.

Tarragon (*Artemisia dracunculus*). Perennial. Start from cuttings taken in early spring. Likes a fertile soil. Divide plants every 4 years to rejuvenate them. The good French tarragon will not come from seed.

Thyme (*Thymus vulgare*). Perennial. Seeds are very small. Start in flats indoors. Germination takes 2 to 3 weeks; 70°F. required. Plants will spread and can be divided. Very easy to propagate by layering. Attracts bees.

Winter savory (*Satureja montana*). Perennial. Propagate by layering.

Woad (*Isabis tinctoria*). Biennial. May be started from seed in the spring; will self-sow readily.

Wormwood (*Artemisia absinthum*). Perennial. Propagate by cuttings, root division in autumn, or by seeds sown in the garden in fall. Seeds may also be started in the winter in flats for spring transplanting. Germination is fast.

Yarrow (*Achillea millefolium*). Perennial. Start seeds indoors in March. Sow seeds on top of fine soil. Plants will spread and may self-sow. Valuable pest-repellent herb.

House Plants and Annuals for Indoors

Acacia (*Acacia baileyana*). Also called cootamunda wattle. Stem cuttings. Seeds.

African violet (*Saintpaulia* spp.). Take leaf cuttings, preferably in March, jerking well-matured leaves from center of plant. Leave 1½- to 2-inch petiole, which can be cut back if it rots. Use 1 inch of gravel for drainage and 2 inches of rooting medium which has been moistened and drained. Do not let leaves touch it. Glass or plastic cover. Keep in good light, not sun, at 65° to 70°F. If beads of water form on cover, remove for 1 hour. If fungus, remove cover and water a little. Seeds, very slow; root division; or crown division when plant not in bloom. Dry divisions 12 hours before planting.

Aluminum plant (*Pilea cadierei*). Also called artillery plant. Stem cuttings of new shoots. Seeds.

Amaryllis (*Hippeastrum hybrids*). Offsets; include some root when separating. Will bloom in 3 years. Seeds: plant soon after ripening, 1 inch apart; soak well once, then keep moist. When leaves are 3 inches, pot up, moving from 3- to 4- and 6-inch pots as plants grow. Keep upper part of bulb above ground. Do not rest plants until after they flower.

Angel-lily (*Crinum moorei*). Offsets.

Anthurium (*Anthurium andreanum*). Division; very fragile roots. Seeds safer.

Apostle plant (*Neomarica gracilis*). Division. Root the plantlets from the flower shoots.

Aralia (*Fatsia japonica*). Stem cuttings in spring. Seeds.

Arrowhead (*Syngonium podophyllum*). Stem cuttings; will live a year in water.

Asparagus fern (*A. plumosus* and *A. sprengeri*). Division; use heavy clippers. Seeds slow; start in January.

Avocado pear (*Persea gratissima*). Suspend large seed on toothpicks in glass of water, root end barely touching the water.

Azalea (*Rhododendron indicum*). Stem cuttings of well-established new shoots. Seeds.

Baby's tears (*Helxine soleiroli*). Stem cuttings.

Bead plant (*Nertera granadensis*). Seeds. Cuttings root easily.

Begonia spp. For *Begonia imperialis*, divide rhizomes. For *B. rex*, stem cuttings, spring or summer. Leaf cuttings. *B. semperflorens*: stem cuttings, spring or summer. Seeds germinate in 10 to 40 days depending on light and heat. Start January. Do not cover the tiny seeds.

Bellflower (*Campanula pyramidalis, C. isophylla, C. fragilis*). Seeds. May be grown as annuals.

Bird-of-paradise (*Strelitzia reginae*). Divide tubers when roots push out of pot. Seeds in sand and peat.

Black-eyed Susan (*Thunbergia alata*). Tropical variety. Plant seeds. Stem cuttings.

Blue dawn flower (*Ipomea learii*). Stem cuttings. Soak seeds and plant in large pot.

Blue lily-of-the-Nile (*Agapanthus africanus*). Root division when repotting in spring. Seeds.

Boxwood (*Buxus mycrophylla japonica*). Softwood cuttings in spring or semi-hardwood in fall, rooted in cool greenhouse or cold frame.

Bromelia family (*Bromeliaceae*). Sever offshoots. Seeds.

Browallia (*Browallia speciosa*). Seeds in rather cool environment. Pinch seedlings.

Cactus family (*Cactaceae*). Seeds, in sand and gravel; stem cuttings, take 2 joints. Keep really dry.

Caladium spp. Division, late winter or spring. Store tubers when dormant in slightly moist vermiculite; plant in spring.

Camellia (*Camellia japonica*). Stem cuttings, at end of growing period. Use bottom heat. (May take 3 months to strike roots.) Division. Seeds.

Cape primrose (*Streptocarpus rexii*). Seeds, sown February to April. Division. Leaf cuttings.

Cast-iron plant (*Aspidistra elatior*). Report and divide in March.

Chenille plant (*Acalypha hispida*). Stem cuttings in fall.

Chilean bellflower (*Lapageria rosea*). Seeds.

Chinese evergreen (*Aglaonema modestrum*). Stem cuttings. Division. Air layering.

Christmas cactus (*Schlumbergera bridgesii*). Stem cuttings. Allow to dry until they callus. Water very little until they root.

Christmas pepper (*Capsicum annuum*). Seeds, sown in February or March. Keep warm; thin promptly.

Cineraria (*Senecio cruentus*). Seeds, sown in light sandy soil in spring. Keep warm.

Citrus spp. Cuttings of half-ripened wood. Seeds.

Coffee tree (*Coffea arabica*). Stem cuttings.

Coleus (*Coleus blumei*). Stem cuttings. Will root in water. Easy and quick (3 weeks).

Cootamunda wattle—see Acacia

Croton (*Copiaeum variegatum pictum*). Stem cuttings, leaf cuttings.

Crown of thorns (*Euphorbia milii, E. splendens*). Stem cuttings.

Cyclamen (*Cyclamen persicum*). Seeds. Keep fairly cool. Blooms in 18 months.

Draecena (*Draecena godseffiana*). Stem cuttings. Air layering. Seeds.

Dumbcane (*Dieffenbachia* spp.). Stem cuttings; use bottom heat. Division. Leaf cuttings. Air Layering.

Ebony or orchid tree (*Bauhinia* spp.). Seeds; keep sterile, they rot easily. Cuttings need heat and moisture to root. Remove leaves.

Ebony jacaranda (*Jacaranda mimosifolia*). Seeds will germinate quickly. Cuttings somewhat difficult; use half-ripe shoots, spring or summer.

English ivy (*Helix hedera*). Stem cuttings.

False aralia (*Dizygotheca elegantissima*). Stem cuttings.

Ferns

> **Bear's-foot fern** (*Humata tyermannii*). Named for the white-scaled rhizomes; divide these for propagation.

They push out over the rim of the pot and are easy to cut or layer. May be slow to root; put in plastic bag for a while.

Bear's-paw fern (*Aglaomorpha meyenianum*). Division. Spores. Grow on damp humus in a clay pot or basket lined with fiber.

Bird's nest fern (*Asplenium nidus*). Division. Spores. Will be small in a dry house; large if humidity high.

Boston fern (*Nephrolepsis exaltate bostoniensis*). Divide long runners, or remove new plants at nodes. Spores.

Golden polypody (*Polypodium aureum*). Divide creeping rhizomes. Spores.

Japanese holly fern (*Cyrtomium falcatum*). Spores. Grow on woods soil with large pebbles and sand.

Leather fern (*Rumohra adiantiformis*). Division. Spores.

Maidenhair fern (*Adiantum cuneatum*). Division. Spores.

Ribbon brake (*Pteris cretica* var. *albolineata*). Divide rhizomes. Spores.

Rosy maidenhair (*Adiantum hispidulum*). Propagation by division. Spores. Use soil with broken oyster shells. Keep always moist.

Silver brake (*Pteris quadriaurita* var. *argyraea*). Easy to propagate from spores; self-propagates on own or nearby pots.

Southern maidenhair fern (*Adiantum capillus-veneris*). Divide and repot often.

Spear-leaved fern (*Ptryopteris pedata* var. *palmata*). Divide rhizomes. Spores.

Staghorn fern (*Platycerium vassei*). Propagate by spores.

Sword fern (*Nephrolepsis exaltata*). Divide runners. Use plenty of broken crock, and pebbly soil. Spores.

Trembling brake (*Pteris tremula*). Divide crowns. Self-sown spores.

Victorian brake (*Pteris ensiformis Victoriae*). Rhizome division. Spores.

Figs (*Ficus* spp.). Stem cuttings. Air layering.

Firecracker plant (*Crossandra infundibuliformis*). Take stem cuttings from this one.

Fishbone plant (*Maranta massangeana*). Root cuttings in spring.

Flowering maple (*Abutilon hybridum*). Stem cuttings of young wood, spring or summer. Air layering.

Fuchsia (*Fuchsia* spp.). Stem cuttings, using young shoots. Seeds, started in the light.

Gardenia (*Gardenia jasminoides*). Seeds. Stem cuttings with a heel. Softwood in summer; hardwood in winter.

Geranium (*Pelargonium* spp.). Stem cuttings, late summer to late fall. Keep plants dry for 3 weeks before taking cuttings. Use 3- or 4-inch lengths, in sandy soil, with half of leaves cut off. Keep shaded while roots forming, about 2 or 3 weeks. Then move to sun. Water from below, just enough so stems do not shrivel. (If stems succulent, expose first to air for callus to form.) When 5 inches tall, pinch back and use tips for more cuttings. Seeds.

Gesneriad family (*Gesneriaceae*). Leaf cuttings. Stem cuttings in water, and pot up when roots ½ inch long.

Glorybower (*Clerodendrum thomsonae*). Stem cuttings, using half-ripe wood.

Gloxinia (*Sinningia speciosa*). Seeds, do not cover. Keep warm. Divide tubers to restart in February.

Haworthia spp. Stem cuttings. Divide offsets.

Impatiens spp. Also called Patience. Stem cuttings, spring or summer. Use tips. Will root in water. Very tiny seeds; do not cover.

Ivy (*Hedera* spp.). Cuttings very easy to root.

Jade plant (*Crassula argentea*). Leaf cuttings. Seeds.

Jerusalem-cherry (*Solanum pseudocapsicum*). Grow as annual from seed. Cuttings.

Kafir-lily (*Clivia miniata*). Repot every 5 years; divide then.

Kalanchoe spp. Stem cuttings; dry for 24 hours to form callus. Seeds, especially of *K. blossfeldiana*. Division of plantlets.

Kangaroo vine (*Cissus antarctica*). Stem cuttings.

Lantana (*Lantana camara*). Stem cuttings. Seeds.

Lipstick plant (*Aeschynanthus pulcher*). Use prunings after flowering.

Magic plant (*Achimenes* spp.). Also called nut-orchid. Divide rhizomes in January. Keep cool. Seeds: press into planting mixture. Pinch seedlings.

Medicine plant (*Aloe vera*). Offsets, root suckers. Dry for 24 hours before replanting. Do not overwater.

Mimosa (*Acacia* spp.). Seeds; scarify. Seedlings are heavy feeders.

Monstera (*Monstera deliciosa*). Also called Swiss cheese plant. Stem cuttings. Layering; air layering.

Morning glory (*Ipomoea tricolor*). Seeds; need soaking.

Nasturtium (*Tropaeolum* spp.). Seeds. Stem cuttings, especially of viny varieties.

Nerve plant (*Fittonia argyroneura*). Stem cuttings.

Norfolk Island pine (*Araucaria excelsa*). Seeds. Air layering. Tip cutting.

Nut-orchid—see Magic plant

Oleander (*Nerium oleander*). Also called rose bay. Stem cuttings. Often will root in water.

Orchid family (*Orchidaciae*). For those which form pseudobulbs, take rhizome cuttings or backbulbs from previous year's growth: 1 to 3 bulbs per division. Plant in thick, dense sphagnum moss.

Palm (*Chamaerops mediterranea*). Seeds. Use high bottom heat. Very slow.

Passion flower (*Passiflora caerula*). Seeds in not-too-rich soil. Layering.

Patience plant—see Impatiens

Pepper elder (*Peperomia* spp.). Seeds. Stem cuttings. Leaf cuttings.

Periwinkle (*Vinca major*). Division. Layering. Stem cuttings. Seeds.

Petunia (*Petunia* x *hybrida*). Seeds are very fine; mix with sand. Cuttings.

Philodendron spp. Stem tip cuttings.

Pickaback or piggyback plant (*Tolmiea menziesii*). Leaf cuttings. Take plantlets from mature leaves; can be layered.

Pineapple (*Ananas comosus*). Cut crown off, leaving 1 inch of fruit. Expose to air for 48 hours; place on sand; water and fertilize from above.

Pittosporum spp. Stem cuttings. Seeds.

Poinsettia (*Euphorbia pulcherrima*). Cut back to 4 inches in April, let new shoots grow, then use these for stem cuttings. Keep at 70°F.

Poor man's orchid (*Schizanthus pinnatus*). Seeds. Keep cool to prevent legginess. Transfer the plant to 70°F. to induce flowering.

Prayer plant (*Maranta leuconeura*). Root division when repotting in spring. Keep warm and moist. Leaf cuttings.

Prickly pear (*Opuntia* spp.). Cuttings. Grafting.

Primrose (*Primula malacoides*). Divide when repotting in fall. Seeds. Use 65°F. bottom heat.

Pussy ears or pothos (*Scindapsus aureus, S. pictus*). Stem cuttings. Aerial root cuttings. Seeds.

Pussy ears (*Cyanotis somaliensis*). Stem cuttings. Division.

Rex begonia—see Begonia

Rose bay—see Oleander

Rubber plant—see Figs

Screw pine (*Pandanus veitchii*). Divide suckers and offsets from aerial roots.

Shamrock (*Oxalis* spp.). Seeds. Divide rosettes.

Shrimp plant (*Beloperone guttata*). Stem cuttings, from prunings.

Silk oak (*Grevillea robusta*). Seeds. Very easy.

Slipper plant (*Calceolaria multiflora, C. integrifolia*). Seeds. Start in June for winter flowering.

Snake plant (*Sansevieria trifasciata, S. t. laurentii*). Offsets. Leaf cuttings. Use offsets or root division to avoid color change.

Southern yew (*Podocarpus macrophylla*). Use prunings for cuttings.

Spathe flower (*Spathiphyllum floribundum*). Divide rootstock at time of yearly repotting.

Spider fern (*Pteris cretica*). Divide runners. Spores.

Spider plant (*Chlorophytum comosum*). Divide plantlets from mother plant. Seeds.

Spiderwort (*Tradescantia blossfeldiana*). Stem cuttings in spring; very easy to root.

Stonecrop (*Sedum* spp.). Stem cuttings. Seeds.

String of hearts (*Ceropegia woodii*). Stem cuttings.

Swedish ivy (*Plectranthus australis*). Stem cuttings; very easy.

Sweet olive (*Osmanthus fragrans*). Stem cuttings; take in summer.

Swiss cheese plant—see Monstera

Temple bells (*Smithiantha* or *Naegelia* spp.). Division of scaly rhizomes. Stem cuttings.

Ti plant (*Cordyline terminalis* 'Ti'). Air layering. Tip cuttings. Stem sections.

Tree ivy (*Fatshedera lizei*). Stem cuttings; use young growth.

Trileaf wonder (*Syngonium podophyllum*). Stem cuttings; very easy.

Umbrella plant (*Cyperus alternifolius*). Tip cuttings. Crown division. Division of runners. Seeds.

Umbrella tree (*Schefflera actinophylla*). Air layering. Stem cuttings. Seeds.

Velvet plant (*Gynura aurantiaca*). Use prunings for cuttings.

Verbena spp. Division, spring or fall. Seeds.

Wandering jew (*Tradescantia* spp.). Tip cuttings. *Zebrina pendula*, same.

Wax plant (*Hoya carnosa*). Stem cuttings; best taken in spring.

Zebra plant (*Aphelandra* spp.). Stem cuttings.

Bibliography

Birdseye, Clarence and Eleanor G. *Growing Woodland Plants*. London: Oxford University Press, 1972. (Paperback)

DeBoer, Roy H. "The Effects of Light on Plants." *Horticulture,* July 1968, p. 35.

Esau, Katherine. *Plant Anatomy*. NY: Wiley, 1953.

Flemer, William III. *Nature's Guide to Successful Gardening and Landscaping*. NY: Thomas Y. Crowell Co., 1972.

Foster, Catharine Osgood. *The Organic Gardener*. NY: Knopf, 1972.

----------. *Organic Flower Gardening*. Emmaus, PA: Rodale Press, Inc., 1975.

Free, Montague. *Plant Propagation in Pictures*. Garden City, NY: Doubleday Publishing Co., 1957.

Gardening with Native Plants. Brooklyn, NY: Brooklyn Botanic Garden.

Garrett, Stephen. *Soil Fungi and Soil Fertility*. Elmsport, NY: Pergamon, 1969.

Graf, A.B. *Exotica*. 3rd ed. Rutherford, NJ: Roehrs Co., 1970.

Handbook on Houseplants. Brooklyn, NY: Brooklyn Botanic Garden.

Handbook on Propagation. Brooklyn, NY: Brooklyn Botanic Garden.

Haring, Elda. *The Complete Book of Growing Plants from Seed*. NY: Hawthorn Books, Inc., 1967.

Hartmann, Hudson T. and Kester, Dale E. *Plant Propagation: Principles and Practices*. 3rd ed. Englewood Cliffs, NY: Prentice-Hall, 1975.

Hillman, W.S. and Salisbury, F.B. *The Biology of Flowering*. NY: Natural History Press, 1969.

Hills, Lawrence D. *The Propagation of Alpines*. London: Faber and Faber, 1959.

Hottes, Alfred C. *The Book of Perennials*. NY: De La Mare, 1947.

Janick, Jules and Moore, James N., eds. *Advances in Fruit Breeding*. West Lafayette, IN: Purdue University Press, 1975.

Loomis, W.E. *Growth and Differentiation in Plants*. Ames, Iowa: Iowa State College Press, 1958.

Masefield, G.B. et al. *The Oxford Book of Food Plants*. London: Oxford University Press, 1969.

Miles, Bebe. *Bluebells and Bittersweet: Gardening with Native Plants*. NY: Van Nostrand, Reinhold, 1969.

Mitchell, J.W. *Plant-Growth Regulators*. Washington: Government Printing Office, 1942.

Northen, Rebecca T. and Henry T. *The Secret of the Green Thumb*. NY: Ronald Press, 1954.

Ortloff, Stuart and Raymore, Henry. *A Book about Soils for the Home Gardener*. NY: Barrows, 1962.

Poincelot, Raymond P. *Gardening Indoors with House Plants*. Emmaus, PA: Rodale Press, Inc., 1974.

Rockwell, F.F. and Grayson, E.C. *The Complete Book of Annuals*. NY: Doubleday Publishing Co., 1955.

Rodale, J. I. and staff. *The Encylopedia of Organic Gardening*. Emmaus, PA: Rodale Press, Inc., 1971. (Articles on plant propagation.)

----------. *How to Landscape Your Own Home*. Emmaus, PA: Rodale Press, Inc., 1963. (Sections on propagation.)

Simmons, Adelma Grenier. *Herbs to Grow Indoors*. NY: Van Nostrand, Reinhold, 1972.

Steffek, Edwin F. *How to Know and Grow Wild Flowers*. NY: Crown, 1954.

Taylor, Kathryn S. and Hamblin, Stephen F. *Handbook of Wild Flower Cultivation*. NY: Macmillan, 1963.

Taylor, Norman. *Wild Flower Gardening*. Princeton and NY: Van Nostrand, 1955.

United States Department of Agriculture. *Better Plants and Animals: Yearbook of Agriculture*. Washington: Government Printing Office, 1937.

----------. *Seeds: Yearbook of Agriculture*. Washington: Government Printing Office, 1961.

Van der Veen, R. and Meijer, G. *Light and Plant Growth*. NY: Macmillan, 1959.

Wilkins, M.B., ed. *The Physiology of Plant Growth and Development*. NY: McGraw-Hill, 1969.

Wyman, Donald. *Wyman's Gardening Encyclopedia*. NY: Macmillan, 1974.

Yepsen, Roger B., Jr., ed. *Organic Plant Protection*. Emmaus, PA: Rodale Press, Inc., 1976.

Other suggested reading:

Briggs, F.H. and Knowles, P. *Introduction to Plant Breeding*. NY: Reinhold, 1967.

Crocker, W. *Growth of Plants*. NY: Reinhold, 1948.

Frey, K.J. *Plant Breeding*. Ames, Iowa: Iowa State College Press, 1966.

Galston, A.W. *Control Mechanisms in Plant Development*. Englewood Cliffs, NJ: Prentice-Hall, 1970.

Garner, R.J. *The Grafter's Handbook*. London: Faber and Faber, 1958.

Kains, M.G. and McQuesten, L.M. *Propagation of Plants*. NY: Orange Judd, 1916 and 1938.

Kramer, J. *How to Grow African Violets*. Menlo Park, CA: Lane, 1971.

Leach, D.G. *Rhododendrons of the World and How to Grow Them*. NY: Charles Scribner's & Sons, 1961.

Leopold, Aldo C. *Plant Growth and Development*. NY: McGraw-Hill, 1964.

Mahlstede, J.P. and Haber, E.S. *Plant Propagation*. NY: Wiley, 1957.

Mayer, A.M. *The Germination of Seeds*. NY: Macmillan, 1963.

Rees, A.R. *The Growth of Bulbs*. London: Academic Press, 1972.

Rockwell, F.F., Grayson, E.C., and de Graff, J. *The Complete Book of Lilies*. Garden City, NY: Doubleday Publishing Co., 1961.

Seeds of Woody Plants in the United States. USDA Handbook No. 450. Washington: Forest Service, United States Department of Agriculture, 1974.

Tukey, H.B. *Dwarfed Fruit Trees*. NY: Macmillan, 1964.

Way, R.D., Dennis, F.G., and Gilmer, R.M. *Propagating Fruit Trees in New York*. New York Agricultural Experiment Station Bulletin No. 817, 1967.

Wells, J.S. *Plant Propagation Practices*. NY: Macmillan, 1955.

Plant Index

297

avocado (*Persea americana*), 164
avocado pear (*Persea gratissima*), 283
azalea (*Rhododendron indicum*), 283
Azalea spp., 10, 19, 20, 73, 192, 193
Aztec lily (*Sprekelia formosissima*), 215

B

baby blue eyes (*Nemophila insignis* or *N. menziesii*), 215
baby's breath, annual (*Gypsophila elegans*), 215
baby's breath, perennial (*Gypsophila paniculata*), 215
baby's tears (*Helxine soleiroli*), 283
bachelor's-button, annual (*Centaurea cyanus*), 215, 224
bachelor's-button, perennial (*Centaurea dealbata*), 215
balloon flower (*Platycodon grandiflorum*), 216
balsam, 216
balsam fir (*Abies balsamea*), 197
banana, 50, 51
baneberry (Actaea alba or A. *rubra*), 247
Baptisia australis (false indigo or wild blue indigo), 218, 252
barberry (Berberis spp.) 193
barren strawberry (*Waldsteinia fragarioides*), 247
basil (*Ocimum basilicum*), 275
basket flower (*Centaurea americana*), or *Hymenocallis calathina*), 216, 241
basket of gold (*Alyssum saxatile*), 214
basswood (*Tilia americana*), 202
Bauhinia spp. (ebony or orchid tree), 285
bay (*Laurus nobilis*), 276
bayberry (*Myrica pensylvanica*), 193
beach wormwood (*Artemisia stelleriana*), 224, 247
bead fern (*Onoclea sensibilis*), 265
bead plant (*Nertera granadensis*), 283
beans, bush (*Phaseolus nanus*), 268
beans, climbing (*Phaseolus vulgaris*), 268
bearberry (*Arctostaphylos uva-ursi*), 247
beard tongue (*Penstemon* spp.), 216, 256
bearded iris, 230
bear's breech (*Acanthus mollis*), 216, 248

bear's-foot fern (*Humata tyermanni*), 285
bear's-paw fern (*Aglaomorpha meyeniaum*), 286
bee balm (*Monarda didyma*), 248, 276
beech (*Fagus* spp.), 193
beets (*Beta vulgaris*), 268
Begonia spp. (begonia), 47, 48, 131, 216, 244, 283, 291
Begonia imperialis, 283
Begonia rex, 283, 291
Begonia semperflorens, 283
Begonia × tuberhybrida (tuberous begonia), 216, 244
bell poppy (*Papaver macrostomum*), 237
bellflower (*Adenphora communis, Campanula pyramidalis, C. isophylla,* or *C. fragilis*), 216, 248, 283
bells-of-Ireland (*Molucella laevis*), 217
Beloperone guttata (shrimp plant), 291
Berberis spp. (barberry), 193
bergamot (*Monarda didyma*), 248, 276
bergenia (*Bergenia cordifolia*), 217
Bergenia cordifolia (bergenia or saxifrage), 217
berry fern (*Cystopteris bulbifera*), 249, 263
Beta vulgaris (beets), 268
Beta vulgaria cicla (Swiss chard), 270
betony (*Stachys lanata*), 217
Betula spp. (birch), 193
Betula nigra (river birch), 193
big quaking grass (*Briza maxima*), 229
birch (*Betula* spp.), 193
bird-of-paradise (*Strelitzia reginae*), 283
bird's nest fern (*Asplenium nidus*), 286
bishop's cap (*Mitella diphylla*), 248
bitterroot lewisia (*Lewisia rediviva*), 248
bittersweet (*Celastrus* spp.), 193
black apricot (*Prunus dasycarpa*), 162
black cohosh (*Cimicifuga racemosa*), 218, 248
black oak (*Quercus velutina* et al.), 204
black raspberry (*Rubus occidentalis*), 186

black walnut (*Juglans nigra*), 190
blackberry (*Rubus allegheniensis, R. nigrobaccus* et al.), 165, 166
blackcap (*Rubus occidentalis*), 186
blackeyed heath (*Erica melanthera*), 200
black-eyed susan (*Rudbeckia hirta* or *Thunbergia alata*), 248, 283
blackstemmed spleenwort (*Asplenium resiliens*), , 263
blanket flower (*Gaillardia* spp.), 217
bleeding heart (*Dicentra spectabilis*), 217
bleeding heart, plumy (*Dicentra eximia*), 248
blood-lily (*Haemanthus multiflorus*), 218
bloodroot (*Sanguinaria canadensis*), 248
blue cohosh (*Caulophyllum thalictroides*), 249
blue dawn flower (*Ipomea learii*), 283
blue gum (*Eucalyptus globulus*), 197
blue indigo (*Baptisia australis*), 218, 252
blue lily-of-the-nile (*Agapanthus africanus*), 23
blue lobelia (*Lobelia erinus*), 232
blue sage (*Salvia azurea*), 258
blue wings (*Torenia fournieri*), 218
bluebell (*Campanula rotundifolia*), 248
blueberry (*Vaccinium cyanococcus* et al.), 126, 127
blue-dicks (*Brodiaea capitata*), 218
bluestar bellflower (*Campanula ramosissima*), 217
bluet (*Houstonia caerulea*), 249
bocconia (*Macleaya cordata*), 237
Boltonia asteroides (false starwort), 225
boneset (*Eupatorium perfoliatum*), 249, 276
borage (*Borago officinalis*), 276
Borago officinalis (borage), 276
Boston fern (*Nephrolepsis exaltate bostoniensis*), 263, 286
bottle gentian (*Gentiana andrewsi*), 252
bottlebrush buckeye, 194
bottlebrush tree (*Callistemon lanceolatus*), 193
bouncing bet (*Saponaria officinalis*), 218
box (*Buxus sempervirens* or *B. microphylla*), 193

NOTE: Plants not listed here may be found in Part II of the text, beginning on page 160. See also the index of plant names for Latin and common names.

layering and, of perennials, 64
trench, 56, 57
mound layering, 57, 58, 64
mulch, cuttings and, 35, 41
pest control and, 22

N

naphthaleneacetic acid (NAA), 28
narcissis, 77
natural layering. *See also* offsets, runners, *and* stolons.
types of, 60–62
nectar, 119, 120
nematodes, 5, 9, 22
marigolds and, 22
nitrogen, 4, 7
node, 33
fern, 81
inter-, 36
stem cuttings and, 35
nutrients, 3–5, 42
nylon stockings; use of, 3

O

offsets, division and, 67
layering and, natural, 61, 62
wild flower, 145
oxygen, in propagating cases, 8, 9
role of, in grafting, 89, 90

P

pasteurized soil, 5, 6, 15
pathogens, 2, 5, 6
peach. *See also* fruit trees *and* Part II.
artificial light and, 18
pollination of, 124, 125

peat moss, acid, 43
grafting and, 93, 103
heat and, 5
layering and, simple, 55
Michigan, 43
in starting mixtures, 2, 34, 43, 44, 147
as transition medium, 60
peat pots. *See also* clay pots.
as starters, 15
types of, 11
for wild flower transplants, 147
pegging, carnations and, 63
layering and, continuous, 56
trench, 57
leaf cuttings and, 47
peony, 69, 70. *See also* Part II.
perennials, division of, techniques for, 68–71
layering of, 63, 64
root cuttings of, 46, 47
seedlings of, light for, 17
perlite, 6
in starting mixtures, 2, 44
pests. *See also name of pest; e.g.,* aphids.
animals as, 21–24
indoor, 20, 21
insects as, 21–24
outdoor, 19, 20
petiole, budding of, in fall, 101
bud sticks of, grafting of, 95
leaf-bud cuttings of, 27, 49
leaf cuttings of, 48
phloem cells, 26, 27
behavior of, in grafting, 87, 88
layering and, 62
phlox. *See also* Part II.
division of, 69, 70

of seedlings, 153
   wild flower, 147
trees. *See also* Part II.
   grafting of, 85, 104–115
    passim
   root cuttings and, 47
   root pruning and, 75, 76
   scion, 103
   seeds of, 134
   trench layering and, 56, 57
trench layering, 39, 56, 57
tubers, 46, 82
tulip, 77. *See also* Part II.

U

underground stems, 145

V

variegation, 47. *See also* division.
vegetable plants, 141, 267. *See also* Part II.
vegetative buds. *See* leaf-bud cuttings.
vegetative propagation. *See* asexual propagation *and* cuttings.
veneer graft. *See* side grafting.
vermiculite, 6
   in starting mixtures, 2, 15, 34, 39, 43
   in transition mediums, 60
vernalization, of seeds, 119, 120, 143
vinegar, 21

vines, 56, 57

W

Wardian cases, 12
water, grafting and, 89, 90, 93, 94
   hardwood cuttings and, 39
   mist-house and, 15
   role of, in propagation, 8, 9
   sterilization of, 14
water plants, 71, 72. *See also* Part II.
water sprouts, 49, 50
wax, 91, 92
whip grafting, 104, 105
whitewash, 93
wigwams, 15, 16
wild flowers. *See also* Part II.
   crown division of, 72, 73
   layering of, 63
   seeds of, 145–148
   soil for, 145, 146
   starting mixture for, 147
wind-pollinated plants, 13
worms, 5
wounding, air layering and, 59
   corms and, 83
   layering and, simple, 55
   role of, in grafting, 86–88
   root cuttings and, 44, 45
   scooping and, 78, 79

X

xylem, 26, 27
   role of, in grafting, 87, 88